Adventures in Agriculture

Adventures in Agriculture

Volume 1

A Collection of Humorous, Reality-Inspired Stories

BY

Audra Brown

First Printing, 2016

ISBN 978-1-944256-01-2

Adventures in Agriculture
P.O. Box 5
Floyd, NM 88118
www.AinAg.audra-brown.com

Ordering Information:

Individual or quantity sales. Special discounts are available on quantity purchases of more than five copies. For details, contact the publisher via the info above.

Printed in the United States of America

First Edition

1   2   3   4   5   6   7   8   9   20   19   18   17   16

More from Audra Brown

Check out her new columns online and in the
newspaper each week.

Go to:
www.ainag.audra-brown.com
for links, archives, enhanced content.

Also stay in touch by following her at:
www.facebook.com/ToughTarget

Upcoming releases:

Tough Target (A thrilling novel of pseudo-realistic
adventure, 2016)

Fire for Hire (A supernatural western, 2016)

Sand Script - A collection of short fiction (2016)

Adventures in Agriculture - Vol 2 (2017)

CONTENTS

"Agriculture is the noblest of all alchemy; for it turns earth, and even manure, into gold..."

- Paul Chatfield

FOREWORD

Audra Brown may be the closest thing to a Renaissance woman I have ever met.

Writer. Musician. Farmer. Rancher. Black belt. Taekwondo world champion. Mechanic. Sharpshooter. Computer geek. Restorer of old cars. She was the kid who would save up money so she could buy---and I'm not kidding here---25 boxes of books from the used book sale at the local library to fuel a deep and vivid imagination through a winter in the country.

But asked to describe herself, Audra will say she is and always will be a farm kid. The adventures in this book all happened---or at least that's her story, and if I know her, she's sticking with it.

Audra is a gifted storyteller with an authentic voice seasoned with a splash of wit. When she tells you that "rain, oiled leather, and the smell of freshly turned earth have to rank as some of the best smells there are," you trust her. Whether she's bottle-feeding a newborn calf, busting ice on a tank, or heading to town for tractor parts, she's got room in her pickup for one more, and that would be our seat.

Pour that second cup of coffee into a to-go cup and come on.

Daylight's wasting.

> Betty Williamson
> Pep, New Mexico
> April 12, 2016

PREFACE

The profession of agriculture is both one of the oldest and one of the most advanced. Sometimes the fact that it's been around for thousands of years tricks us into thinking that it's more or less the same.

Sometimes it is. Sometimes it ain't.

There is one particular myth, perpetuated by popular culture in movies, TV, and other entertaining media, that these stories might argue with a bit. It's not nearly as dull and boring as you'd hope.

It's an adventure.

And I love sharing it with readers. I get to do it every week thanks to editor David Stevens, and to Betty Williamson, who recommended that he try me out as a columnist.

I am also indebted to my family. They are the reason I have these stories to tell and frequently the inspiration as well. They are both the causes and the victims of my success.

The biggest thanks, though, is to the readers. You are the best. Every time I get a letter, meet one of you in town, or just hear from a friend that you enjoy reading my columns, it makes my day.

Ya'll keep reading. I'll keep writing.

> Audra Brown
> Bethel, New Mexico
> April 13, 2016

WORK FROM HOME, LIVE AT WORK

The phrase, work-from-home, might be used to describe the life of a farmer. But it is perhaps more accurate to invert the concept and use the term, live-at-work.

The farmer may go home, but he is still at work. His mind is never free of the list of things that must be done. No hour of the day or night is divorced from responsibility.

The farmer sleeps because he must, and eats when it is convenient, but on a normal day, never leaves his work.

It doesn't mean that the work is always close and convenient.

There might be fields and barns and pasture in view of the house, wells that can be checked before breakfast, sprinklers that can be seen from the porch, and cattle that graze where another might have a yard. Completely contiguous property is a lovely thing, but far from the common reality.

Ten or fifteen miles of rutted roads just to check all the sprinklers and pumps and cattle on the home place would be an easy morning. Another fifteen both ways if you need a part that you have to get in town.

There is a lot of mileage to be racked up just around the home place.

Now, onto the property that is two counties over and a hundred miles away. A hundred miles to check the water and a hundred back.

It's a good distance and four hours of road, but it can be a pleasant and relaxing break from the usual, overly busy day.

You can turn the radio up or enjoy the peaceful rumble of road noise and just contemplate the universe.

The time between places is perhaps the closest thing to a break one gets in the ag business.

You have to make the trip, but between the job you leave behind and the one waiting ahead, there is nothing you can do but drive. A simple interval of peace amid the normal chaos of the day.

These long roads are often remote, and the one other vehicle you see is a pickup much like your own.

A hundred miles of lonesome peace between the fields and pastures and the work that is your home.

Audra Brown might be in that one other vehicle you see.

HOLD YOUR HEADER UP HIGH

Just about the time it starts getting good and hot, it's time to get to work.

Now, farmers are plenty busy all year round, but they are especially overworked when its planting time or harvest time.

This part of the world, June is one of those times.

We know it's coming, when all that green wheat starts heading out and then gettin' dry. It's amazing how quick our hot, desert days can turn green leaves into dry straw and red grain.

It always sneaks up on me. It starts to turn and I head out to shake the rodents and dust out of the combine. You change the oil, and make sure the air-conditioner does more than blow hot air on your face. But before you can get the header hooked on and the mechanisms greased, you turn around and the wheat is ready to cut.

I don't know that I've ever started a wheat harvest fully prepared, no matter how on the ball I think I am.

One way or the other, the fun starts.

It's a fact that I consider combining to be one of

my favorite jobs. Sittin' up high in a big glass cab, a panorama of swaying grain spread out below, tunes on the radio, and if I'm lucky, a big, cold cup of iced tea within arms reach.

It's not particularly fast.

You hope it's dadgummed slow.

Good wheat won't cut faster than a mile or so an hour— sometimes more like point-five. Those years, it'll take a whole hour just to make one round back and forth across the field and that's not counting the stop for a bin-full. Those times are sweet.

The rest of the time...

You go just as fast as you can without jamming the header into the dirt. The wheat's so short that you lower the header so that it only scrapes the ground on the high spots, you hoof it along at 3 to 5 mph, and you hope to get a bin-full more than once a day.

This was another one of those years. So here's hoping that the combine will get to run slow with the header high next time.

Audra Brown probably shouldn't write while on the combine, but she does...

MIDNIGHT OIL AND TEN O'CLOCK GREASE

Burn the midnight oil, but don't set anything on fire.

It is a well-established fact that farm work does not adhere to a particular set of hours.

From day to day, the start time, end time, and total time between can vary from a short, eight hours, to a good sixteen or more. Ten or twelve is pretty common. But you know what? Farmers operate on a fixed set of hours more often than you might think.

Wheat harvest is, or at least the combine-driving part, is a 10-1 job.

That gets you out before the sun has had a chance to completely bake everything and make the morning maintenance greasing too unpleasant, and yet hot enough that the wheat stalks are just about dried out from the last night's dew.

Barring encounters with some distant engineer's annoying design features that require you to hang off the side and blindly seat chains on sprockets (or some such breakdown and repair) everything is operational right on time and continues uninterrupted until just after midnight.

Shortly after the day hand on your watch rolls over, the nightly dew starts setting in and as painful as it is to quit...when the wheat starts getting sticky enough to wrap around the header instead of sliding on inside to get threshed, it's time to call it a morning and go catch a few winks.

Find a spot that's either green or bare--terrain uninclined to catch fire--and put the combine in park. Throttle back and disengage the threshing machine.

As you enjoy the silence of nothing but the engine running, set back and listen to one more song on the radio while the motor cools down.

Ice-chest and lunchbox in hand, turn the lights out and the fuel off, climb down the steps, and go home for a cup of coffee and a few hours of sleep.

Same place, same time, next morning.

Audra Brown has used a lot of grease.

BACKWARDS IS THE NEW FORWARDS

Many hours and days and nights have I spent on a big, red machine. There's even a song about it.

It gets quite a bit right. A combine isn't fast. Roading it with the header off, you can get close to 15, but rear-wheel steering at that speed makes things get a little exciting sometimes. Ten is more stable, but when the header is on, and its cutting time, five is hoofin' it.

Backing up traffic is amusing. Pull out onto the Bethel Highway with the header on and it's fence to fence. Ain't nobody getting around, and it ain't possible to do anything but keep going where you're going in the combine.

Then there are gates... Very few gates are designed with thirty-foot headers in mind. So, one of the primo skills as a combine-driver, is the ability to squeeze a big machine with a thirty-foot header through holes that are considerably less than thirty feet.

There are two basic methods: the forward twist and the reverse pivot.

If the gap is only a little too small, the forward method works. You simply approach the opening from the side, paralleling the fence as close as possible.

When the header is even with the gate, you start turning into the gate and swing the end of the header around before the rest of it. If performed correctly, the combine gets through the gate and all the fences are still standing. The shortfall of this technique is that you must remain a header-half away from the fence prior to the turn, and since combines don't turn incredibly short, it only shaves a couple feet off.

Theoretically, one only needs a gate that is the width of the distance between the outside of the right tire and the left edge of the header. (I do mean the right tire, because it doesn't have a ladder in the way.) The easiest method to squeeze through such a minimal space is the reverse pivot. Simply back the combine into the gate so that the right tire just clears the right gatepost. Similarly, the left end of the header should be lined up so that it will go just inside the left gatepost. (If not...then you're just gonna have to face the fact that it won't fit.) Just before the right side of the header runs over the fence, you apply the right brake and turn the wheel so that the combine backs around and through.

Easy as that.

Audra Brown only does some things backwards.

THE TROUBLE WITH COMBINES

Wheat harvest. Grease in the morning, run until the stalks start wrapping around the auger in the header, then get a few hours sleep. You can even hope for a shower between sometime after midnight and before its much past daylight. The hours are long and odd, the working schedule is even worse.

One day on, one day off.

The day on, the combine is working. The header turns, the chains are on the sprockets, the belts are in one piece, and nothing is on fire. As long as the A/C is working in the cab, these are the easy days. Nothing to do but sit in a relatively comfy chair, listen to the radio, and pretend you're the captain of the starship *International Harvester*.

The days off are bad. The combine is broke down.

It's not a work day because its not working. But the farmer is, trying to fix the combine and get it back in the field, cutting the wheat.

Working on a combine is never a simple task, and unfortunately, it's not a rare occurrence either. Combines are tall, complicated, disaster-prone pieces of machinery that were designed to do a specific job amazingly well.

Not designed to be easy to work on.

There are a plethora of belts, chains, and other moving parts that have a tendency to break, stretch, and jump off their designated pulleys and sprockets.

Once a replacement is found--either a new part, or something not yet broke that you commandeered off a piece of equipment that doesn't have to be running today--things can get exciting.

One of my favorites is a particular chain that comes from the factory tensioned by a static wooden idler that inevitably wears through, breaks and allows the chain to jump the sprockets or break.

Rethreading this chain (that is very long, and thus reasonably heavy) requires that you climb up in the bin, hang out over the side, and open up a little access door that's not big enough to reach inside and see at the same time. Reaching inside with the one hand that isn't keeping you from falling to the ground, all you have to do is thread the errant chain around three sprockets that you can't see--in the right order. And you better thread it around them the right direction too, or it still doesn't work.

You do it until you get it done, or until you fall off.

Audra Brown has hung off the side of a lot of machinery.

ANOTHER HARVEST OLDER

There once was a kid, who turned a year older.

Friends and family congregated and waved down the combine to stop at the edge of the field. They brought food and presents and a pinata to pummel.

The field was flat, no trees around, so they hung the paper-mache box from the unloading auger.

The big sister got in the cab and made the pinata dance by flipping some switches and swinging the auger around a bit.

The younger kids whacked away with a shovel or some other lever suitable for pinata destruction and eventually candy was collected and dug out of the sand.

I wasn't born last night, but I may have indeed fallen off a truckload of hard red winter wheat.

The combine stops for nothing but parts when it's time to get the wheat out of the field and I'm one of the lucky souls who pretty much always has my birthday in the middle of it.

Many of my birthdays happened on a combine, in the field, and that's okay.

Some years, along with supper, there might be a plate of cake (or hopefully, pie) with a candle or two stuck somewhere.

Being careful about where you light 'em and where you blow 'em out, so as not to set anything on fire unintentionally, it's not a bad way to celebrate.

Some of the best birthday parties are the ones that weren't really planned at all, and that tends to be the way it goes in the field.

A few years back, I got flagged down and pulled over at the end of the field, climbed down from the combine, and found a significant collection of family and friends gathered to sing me the classic song.

Siblings, parents, grandparents, aunts, uncles, cousins and friends, made for a fun and unexpected crowd and a darn good way to remember a birthday.

Audra Brown doesn't always get a birthday party, but when she does, its probably in the field.

SELF-PROPELLED
SELF-DESTRUCTION

The term, "self-propelled," is pretty self-explanatory, and it sounds like a good thing. It has a place. Certain equipment is just not disposed to being configured as a pull-type rig. Grain combine, cotton strippers, and even swathers have some justification.

But there is a darkside.

The problem is, that despite the ease of use when operational, a self-propelled piece of farm equipment is really two pieces of equipment fused into one big, double-the-trouble machine.

There is the machine; the moving parts and tricky ideas that is built to accomplish a specific task. Then there is the motivation and habitation, duties that would normally be the purview of the tractor. Any specialized machine has plenty of breakdowns and things that can and will go wrong. Tractors are no different. But if they are separate, a broke-down tractor can be switched out, and a broken implement doesn't mean a useless tractor.

Self-propelled means more to take care of. The implement maintenance might not change much, because if you need a machine to do that job, you'll have it in some form, self-propelled or not. But do you really need that extra engine? And more tires?

Engines are known to require certain, regularly-scheduled maintenance. One is less likely to neglect caring for the car that drives you to town, the pickup that drives you to the tractor, or the tractor that you drive all the time.

Engines you only see once a year? They are easy to neglect. Batteries go dead, wires get mysteriously disconnected, fuel gets stale At least most of them have a regular schedule. Even if regular means once a year, at least the combine and the swather get predictably repeated care.

Then there are the rigs that really shouldn't be self-propelled. Like the old well-pulling rig, or the old flatbed bobtail, or the old winch truck. There's a theme here. A collection of oddly configured vehicles, no longer suitable for professional service in their assigned trade, are collected by the farmer to handle the odd jobs that require special specialization.

The old one-ton pickup chassis that propels the odd contraption is only worth the rig grafted behind. Its battery is always dead, its doors and windows might work, A/C is one thing it's never heard of, and don't mind the smoke billowing from the hood. If you can't see flames through the holes by your feet, all is well.

Audra Brown didn't have to put the engine out this week, but someone did.

THE BIG LOANER TRACTOR

There once was a tractor, that was big, green, and mean---in more than one sense of the word.

Now, I know what you're thinking, but you're wrong. It wasn't a John Deere. It was a more chartreuse shade of tractor known as a Steiger. Actually, it was "The Steiger." Referring to it without implying a certain amount of excess personality and gravitas is unacceptable.

In case you were wondering, horsepower, torque, reliability, and cold A/C are what you want in a tractor. Personality and a trailer-load of stories is not.

The Steiger's story was an epic.

The short version starts with a big red tractor and chunks of massive planetary gears, quite literally, bringing things to a grinding halt...in the middle of the field. When the planetaries are locked down, the wheels don't turn. When the wheels don't turn the tractor does not move. It gets plowed around and the sprinkler knocks the windows out, and it just sits there until the gears get rebuilt and put back in.

But crops don't wait and the plowing needs to be done. Finding the parts was almost as daunting as putting them together, so, waiting on Big Red was not an option.

Now, there are the kind of tractors that you like to borrow; the kind that you feel a little guilty about using; the kind you hate to have to return.

Then there are the kind that you usually borrow; the kind nobody misses; the ones you would take down to the sale yesterday if you could.

The Steiger had seen better days, but it started and made the mile or two trip to where Big Red had left off. The ground got plowed and the driver got to work on a six-pack--not of something to drink...

The seat was stuck in the maximum reclined position, so, on the straights, the driver had to look between the spokes of the steering wheel, and on the ends, you had do a sit-up in order to reach the gearshift before the turn. The air was in a condition similar to the inside of a sealed, glass-walled oven. The radio had a wire or two loose, so there was no distraction even if it would have been able to play over the vibrato roar that is 400 diesel horses with only a rotted-out muffler in between.

That was a day on the farm. It was only the first of many long days with The Steiger...

Audra Brown wrote a song or two about the Steiger and kinda almost misses it.

A TREAD AWAY FROM DISASTER

The big, mean, lime-green machine that rolled in to sub
for Big Red, had no brakes, nor muffler, nor air.
It came with a story instead.

Things started off alright, glad to be back in the field,
because the plowing needed done. There was plenty of
plowing and no time to waste, but while it was better
than nothing, we weren't going to win any sort of
race. Each day ought to have started with warming up
the tractor, putting an ice-chest behind the seat, and
greasing the plow. The day is supposed to end with
cooling down the tractor, putting the ice chest back in
the pickup, and heading to the house.

A good day running the Steiger started with some parts
and fixing whatever had shut us down the day before.
A typical day ended with another critical breakdown.
I figure it averaged out to something along the lines
of 50/50. An equal amount of time spent plowing and
broke down.

Things like A/C and a new muffler might not have kept
the tractor from running, but a heat-stroked driver
isn't much use either. That being said, about to cause a
catastrophic failure, they were not. That came next.

There were too many repairs to even remember, but

some of the closest calls can't be forgotten. One day started out fine, plowing a sandy, slightly hilly, circle.

The passes down were good, but the passes up the slightly inclined field seemed to be overly taxing for such a tractor. Losing power is a terrible experience, and then it got worse, and worse, and worse, until it was all but useless.

The lesson here is change your air-filter regularly. It had been way too many years since this one had been switched and the outer was so packed with dirt that it was hard to lift. The inner was dirtier than anything I'd want to see, and at any moment, it could have given way and let all that gunk down in the engine. That would have probably been the end of an adventurous tractor.

But it survived another day and the fuel lines failed, and that was a diesel bath. Then there was a u-joint in the drive-line, that, if it had completely busted loose, would have torn up more innards and wrecked it underneath too severely to contemplate. And hopefully, no shrapnel came into the cab and wrecked the driver too. Changed that one just in time.

The story is too long, and probably not as exciting as it seems to me, but it was an adventure, and a glimpse of the truth that is life on the farm.

Audra Brown wrote at least one song about the Steiger.

TRIED AND TRUE...OR NOT.

Agriculture has been around for a few thousand years.

It's one of the oldest industries and arguably the most critical industry relative to human survival. Perhaps it is the gravitas of such long and uninterrupted establishment that leads to the social perception that it is a constant, primitive, nigh unchanging field, with tried and true methods that rarely progress.

Well, eh...I'm trying to think of a constant beyond the arbitrary job description: grow stuff that people need.

Not having much luck.

The degree and pace of innovation adoption does vary from farmer to farmer. But even those who change slow, still come along eventually.

Whether it's parallel-tracking GPS, hydraulic cattle-chutes, smartphones, or bovine-applied antihistamines, technology and science is leveraged to increase efficiency and reliability in order to maintain an effective agriculture operation.

Before we all had a GPS in our car, a farmer was using an early handheld GPS device to layout fencelines. Before we all had a comm device in our pocket, a farmer put radios in all his vehicles, his house, and in his pocket.

And currently, the farmer has a setup to retrofit a vehicle to drive itself.

It's not unsupervised automation yet, but it's pretty amazing and we've had it long enough that we have obsolete parallel-tracking devices, and old GPS-steering rigs that we really ought to update.

Obstacle avoidance and implement/tractor impact avoidance when you turn on the end is still the purview of the human tractor-driver, but on the straight pieces from end to end, the farmer gets to lean back and pay more attention to the implement status and such.

Or, if everything seems to be working like it ought to, the farmer might read up on the tractor's operator's manual to remember how to program a control sequence so that on the ends, there's only one button to push for the implement to raise up out of the ground, the throttle to pull back, and the transmission to downshift a couple of gears for the turn. Then one more button to put it back in the ground, throttle up, and get back to running gear after the turn.

Or the farmer could just pull a smartphone out of his overalls pocket and watch Netflix.

Audra Brown learned how to plow a straight row manually, but much prefers not having to.

THE TRAKTOR IS ON FIRE AND OTHER SAYINGS

On the tractor.

That's where I've spent a lot of my life.

Back and forth and forth and back. Downshift, implement up, turn, implement down, speed back up, drive straight, don't run over any junk or wells...repeat until through. All day long for days on end.

Some might look at it as a monotonous, even redundant kind of job.

But it doesn't have to be.

Before the advent of the tractor cab (a more noteworthy advancement than sliced bread, in my opinion) singing and thinking were definitely the main past times. I'm pretty sure that lots of ideas and more than a few sour notes were hatched before cabs were common. I'll admit that my experience is limited to a few days hauling hay on what is now called an antique.

Post-cab is my area of expertise.

For a certain, finite number of days, I can listen to the radio. Music until you know all the songs, talk-radio until you know all the rhetoric.

You can learn a lot about current events, politics, and if you work late enough, about the Hatmen and the Shadow People. Coast-to-Coast AM starts about midnight and I've gotten more than a few laughs and fiction ideas thanks to that show.

But my favorite thing to do on a tractor is learn something---like how to speak Russian.

I'm not kidding. The tractor is by far the best place I know of to learn a new language. Audio learning kits are an essential accessory for any extended farming assignment. There's no one there to make fun of you while you're pronouncing strange sounds and if you get frustrated, there's plenty of time to listen to that lesson again, and again, until you've got it down pat.

Maybe I should recommend a more useful language than Russian. I'll admit that I haven't had much opportunity to use it other than to talk to my brother where no one can understand. But I yell at the television when the subtitles are wrong, and when I need to, I can tell a Russian important things like, "I want something to eat," "two beers, please," "heck no," and the most critical, "the tractor is on fire."

Audra Brown says, "traktor ne webinya."

TO STEER IS NO SMALL THING

To be articulate is to speak well.

To be articulating is to drive badly.

The common form of steering, used in tractors, cars, and red wagons, is front-wheel steering. Front wheels turn, the vehicle moves in that direction. The front first, the back last, the middle gets there in the meantime. Quite controllable.

Combines are backwards. Rear-wheel steering makes the back end go in the direction opposite you steer, and while a bit more interesting than the other, is still a relatively stable steering mechanism once one becomes accustomed to it.

There are rigs where the steering is accomplished not by changing the direction of wheels, but by changing the speed. Skid-steering frequently forgoes the traditional steering-wheel in favor of dual levers that allows one to intuitively alter direction by some combination of pushing and pulling. Maneuverable, easy-to-learn, and again, controllable.

Sometimes such luxuries are foregone in favor of something more eccentric and prone to fits of unrecoverable overcorrections.

When you need a turning radius that's practically a negative number, but less skidding than a skid-steer, crazy wheels seem as reasonable as they ever get. Self-propelled swathers steer by changing the speed of the front tires, and have crazy-wheels in the back. Even after years of operating the mad machines, unintended 360° spins in the turnrow will inevitably happen.

Precise it is not. Exciting it is.

Then there are those monsters that pull the biggest plows or lift the biggest bucket of dirt. The steering wheel goes right, the front and back do too, and the middle goes the other way. Articulating tractors usually have four pairs of duals with tires that are bigger than many cars. They do most of their work at speeds approaching six miles an hour. They have one point of articulation alone, two when pulling a plow, and quite often, three. Going forward is okay, backing up is an exercise in calculus and trigonometry.

With smaller tires, and a wider domain of operation, a front-end loader only articulates in one place, but one overcorrection at high speed (by that I mean 20-25 mph) will turn a reasonably straight path into a zigzag of increasing severity in only a moment or two. Attempts to correct are futile. Just slow down, recover, and see how long you can hold it before the next articulation attack.

Audra Brown is crazy about crazy wheels.

NOTHING BUT THE TURNS

What might you hear if you were to pass by the tractor that I'm driving back and forth through the field?

A loud rumbling roar of the diesel engine, I suspect.

What you might see is quite another thing altogether. Disregarding what you might be doing driving in the field (unless lunch is involved, there's not really an excuse), depending on the mood and what I remembered to bring with me that particular day, you might see any one of the following incongruous sights:

Reading a book.

Indeed, the advent of self-steering GPS technology for the slow-moving, row-inclined work of driving a tractor gives me the chance to read without the risk of quite so many bobbles in my pass line. As long as my well-trained peripheral vision and internal row clock are working to remind me when to look up and turn, it all works great.

Writing a book.

This one is logistically the most difficult, because, of all things, a place to set a laptop or a seat that sits you in a position to put a laptop in said lap, were not considered important during the design of the tractor.

Playing the mandolin.

This is very specific. There is a reason. I'm not even really a mandolin player. But the cab of a tractor is just not well suited for the practice of most instruments.

Of those that will even fit through the door, a guitar is just too unwieldy and there is no good place to put it when you turn on the ends. A fiddle seems more compatible if you consider the size of the instrument, but the three-dimensional space required to actually play the fiddle is many times larger than the fiddle itself.

Hence, not good.

The mandolin, however, is just right. Small instrument, small playing space, and no one around to hear me play it.

Audra Brown has not run over a sprinkler, a fence, or any livestock...yet

A COWBOY CLIFFHANGER

It was a couple weeks before Christmas, or otherwise well into winter that year...

The point is, it was cold and had been snowing for days, and it sure hadn't gotten anywhere near thawed.

Been on the ranch doing cattle work that had to be done, while all the young'ns spent most of the time in the semi-truck with the engine running, keeping warmer than the rest of the crew.

Except for the occasional excursion to let fly a few well-packed snowballs or to give the new puppy a chance to run around, the truck was the place to be.

Finally the time came to head back the hundred miles to home, and nothing stood in the way except for the angling road up the side of the valley that wasn't too bad except for the hairpin turn and steepest section right at the top.

The pickups were strapped on the semi's flatbed trailer and everybody climbed in the truck. Gonna get back home that night...or so they thought.

Now, as you may know, and in this case, this type of truck transmission has a high range and a low...

...and was stuck in the high position thanks to some ornery rat's teeth finding an air-line to chew on.

But surely sixth-gear would be sufficient?

And maybe it would have been, if not for that darned, slick snow.

The truck turned into the last corner, but didn't make it up. It stalled just before the end and right after there was any chance to back back down.

Stalled, stranded, and praying the airbrakes hold, the truck was stuck and it was getting quite cold.

There's nobody close, nothing for many a mile, but after four hours of freezing, and reaching through the floor, they did eventually get that savior known as low range.

The trick was the ferrel off a Ralgro gun that was luckily in the toolbox in one of the pickups on the trailer (neither of which could be unloaded without falling off a cliff either.)

The truck was a cold, scary place for a while, but everyone and everything made it home in intact, except for the Ralgro gun.

Audra Brown finds the combination of snow and cliffs uncomfortable

TALKING ABOUT THE WEATHER

How's the weather been? Many a conversation has begun with this question and many mistake it for inconsequential small-talk. For one involved in agriculture, it is a serious and important question.

Walking around the fair, running into friends and acquaintances, most of the conversations start with a "Hey, how are you?" and then quickly proceed to the, "Been getting any rain over your way?"

In response to the not-such-a-droughty year we've been having, the more optimistic individual even used the, "How much of this rain you been getting?" in a non-sarcastic fashion.

While those for whom the weather is not such a critical player, a casual meteorological query might be answered with simple responses like, "Fine," "Good," or "Bad."

When one asks of a farmer or rancher, do not be surprised to receive a detailed report of exactly how much rain has been recorded in various locations on that person's land, when that rain came, how much rain is still needed...

...which crops need it, which crops don't, how it fell (quick, slow, gentle, windy), how much the neighbors got or didn't get...

...how many vehicles got stuck in the mud, whether the irrigation was on or off or changed because of the moisture...

...how tall the grass is, how green the grass is, how much dry grass one is expected to be there for the winter...

...how bad it would be if it all caught on fire, how good it smelled, how bad the corrals are (muddy, stinky, and slick)...

...and so on and so forth.

It can literally go on for hours.

If this seems daunting and one wishes to avoid a prolonged discussion of meteorological facts and history, it is possible to abruptly abort an extended conversation by complaining about the rain.

Be warned, however, that depending on the temperament of the conversant and the condition of the crop, the reaction might be a silent stare and reduction of respect, or, it might result in the rain-hater getting punched in the face. One can never be sure.

Audra Brown's advice is to nod and say, "Hope it rains some more soon," rather than risk it.

PRAISE THE LORD AND PASS THE UMBRELLAS

Rain, oiled leather, and the smell of freshly turned earth have to rank as some of the best smells there are.

First off, let's all give a hearty hallelujah. It rained. I mean, it RAINED! And enough do some good, even.

If you're not happy about the blessed moisture, then best keep your mouth shut.

Living in a desert, water from the sky is kinda like magic--wondrous, transformational, and dadgummed rare.

Now, I'm not saying that there ain't--particular inconveniences--caused by a rain. But I'd much rather be stuck in the mud than in the sand--or really bad, in a pile of manure...

This last week, I've had to pull my pickup out of the mud, ford 18th Street, and wade through the particularly nasty muck that is a set of boggy corrals.

But I'm still smiling--and so is the world.

It's amazing how much better the world looks, smells, and sounds after a good rain.

The grass turns green, the air breathes easier and sweeter, and even the birds and the frogs start singing.

If you want proof that farmers are the most ridiculously optimistic people, just wait for a spring rain. Plows and planters start coming out of the woodwork, sowing circles to everything from cotton to corn, milo to hay grazer, and who knows what else.

So go outside, take a deep whiff of that smell, give thanks for the rain, and pray for more.

And if you don't like it, for heavensakes, don't tell anyone. Them's fighting words in the desert.

Audra Brown is singing too.

SO THAT'S A BLIZZARD

Words that represent rarely encountered concepts can start to lose meaning, diluted in the attempt to find a use for them. Before I'd been in the middle of the great "Tire Fire" of 2011, I'd never understood the meaning of the word wildfire, assuming it to describe a fire in the wild rather than a fire that is, in fact, wild.

On the farm, any given day presents a new battle to be fought. From the difficult to the improbable, obstacles and challenges are met, and no matter how seemingly unlikely, one does not waver in the belief that it will be overcome.

"Don't tell me the odds," is a good mantra.

If it can be done, then do it.

This unflinching confidence is very effective, but every now and then, a force comes around that is, in fact, unstoppable.

Nature is the eternal opponent of those who make their living in agriculture. A daily saboteur, its effects can be fought most of the time. But when it decides to unleash its power, it cannot be fought.

I understand fights that I can't win, but fights that cannot even be fought are truly sobering.

Whether the wind is blowing flames across the world or blowing frozen snow, there is little to be done but secure you and yours as best one can and wait for the maelstrom to pass. The damage will be done, mitigated in places where wise preparations were made, but no guarantees, and no certainties exist.

As soon as the assault has passed, we go out to see what is left--if we can get that far. The day after the wildfire, material was removed, not really hampering the exploration of the damage. Blowing snow on the other hand, blocked everything but a bulldozer from getting through in some places.

Damages and impacts are yet to be truly known, but signs and roofs have felt the wind, roads are still drifted over, vehicles are snow-locked, plans are canceled, cattle are dead, and the experience of those two days of survival is surreal and almost too intense to believe.

I'm thankful that we all seem to have made it, and it only lasted as long as it did, but I'll not forget it soon.

I think that now I know the meaning of the word blizzard far better than I did before. Here's hoping I only have to use it to order tasty, blended ice-cream and candy for a long while after this.

Audra Brown is enjoying being warm.

TORTILLAS COME FIRST

The hundred-year blizzard is now in the past (and it better be at least that long until another one like it comes around) but just in case, lessons have been learned regarding preparations for such a storm.

First, last, and without question the most important preparation you can make is to stock up on tortillas sooner rather than later. Tortilla-lines were long and tortilla-buyers were often left disappointed and tortilla-less even after braving the aggressive crowds of tortilla-seekers. I myself was lucky and picked up the last package off the shelf and slipped away quickly before the lady behind me noticed that there were no more.

After you've secured the mandatory flatbread, it's a good idea to look beyond such essentials and consider a few other life-sustaining preparations.

Drinking water is not a bad idea, as power failure is possible anywhere and more certainty than not outside the most urban of areas. The farther you get from concentrated population, the longer it'll stay blacked-out. Without power, the faucet is useless. A few gallons of drinking water will keep you hydrated, and filling up the bathtub or other containers with not-planning-to-drink-water will give you something to fill the toilet bowl and flush a couple times a day.

Make sure you have a couple of good, old-fashioned books to read by flashlight or candle. I haven't had the chance to sit down and read so much in a long old time. Note that candles produce more heat than flashlights and so are noticeably pleasant in the middle of a winter storm.

A little butane camp-stove is something that is definitely going on my list. You don't know how depressing it is to realize that one has no good way to make coffee until you are completely hot-water-less. It also can keep you warm and cook food.

The best thing I did remember to do is gas my pickup, charge my phone, and charge up the portable power-bank for my phone. Electricity, heat, and hot coffee may have left me for 24 hours, but I could call, lob angry avians, and check Facebook. Roughing it, without a doubt.

Audra Brown knows that internet and tortillas are the foundations of life.

METEOROLOGICAL ISSUES

The weather likes to keep everybody guessing and seems like this time of year, it likes to put out a few, nice, warm days before swapping back to cold in the middle of the day and catching me with my coat in the pickup, rather than between me and the chill.

It reminds me of a conundrum that can be observed in the choice of winter-wear on the farm. Cold weather is expected, and maybe some snow at some point every year. And even more, out on the farm, you expect to have to be out in it. One might expect that a good set of winter clothes would be hanging in the closet, and one would be right, but one would be wrong in assuming that those, nice, warm, wind-breaking, waterproof duds were worn when they are needed the most.

Nope, that stuff only comes out once every blue moon when I make it up to the mountains and snap a couple of waxed planks to my feet for a few short, daylight hours.

In contrast, on the long days out in the cold, breaking ice and working on whatever needs work until its too dark to see or too cold to run a pair of pliers.

On those days, the attire is probably a coat that's got a little thickness to it, but its been worn and worked in for so many years that the holes really cut down on the insulation properties.

Your hands get some gloves that don't do much more than cut the wind, and neither are waterproofed. So, when the ice-water in the tank splashes back, things don't dry out in any great hurry.

For the coldest weather, the coveralls come out.

Coveralls are like ski-bibs, only not as warm, or waterproof, or good at breaking the wind. They also have the added bonus of being cut so straight that finding a pair that fits well enough that you don't waddle around like a penguin, is difficult.

The only place I don't usually skimp is warm boots and a good stocking cap. It's silly that my warmest, most pleasant clothes get worn maybe once a year, but for all the rest of time, it's holes, cold fingers, and not being able to step over a six-inch fence.

Audra Brown likes the cold in the summer.

SIMPLE RECIPE FOR WINTER FUN

Winter brings the cold and night comes far too early. It is the time of evenings gathered around a fire---or gathered a safe distance back from a serious diesel heater. The point is to stay comfortably warm until it is actually late enough to go to sleep for the night.

Work doesn't stop, but more of it happens in the barn than on top of a sprinkler. The season also brings a harvest of get-togethers with friends and family. And there, coffee or cocoa in cup, many stories are told and times are remembered.

As we all know, entertainment is in great demand as a kid on the farm and it can be found everywhere. I myself fondly recall days playing with a pallet of leftover bricks. We had our little town of loose stone and a great deal of fun trading odd-shaped or uniquely colored bricks as if they were made of gold. Climbing trees, sheds, or peanut thrashers, and the subsequent thrills of jumping off.

When ice, snow, or at least slushy mud appears, some form of sledding is inevitable. A rope, a vehicle, and some sort of sled-like piece of something brings hours of cold, wet, and muddy fun.

But I heard one the other day that was new to my ears.

If in the dairy business you are, there is no rest from season to season. As with the mail--rain, wind, sleet, or snow--cows must be milked no matter the condition. But fun still finds a way. What can a kid do when the lots ice over, and the milking is done?

Sledding, skiing, and sliding in general, are recognized as reliable sources of recreation. Thus, one might say that all you need for a good time is a slick surface and a form of external motivation. Surfaces such as snow, water, and ice are good sources of slick. Popular forms of motivation are: gravity, motor-vehicles, boats, and...cows.

It can (and has) been done from New Mexico to Michigan, and likely elsewhere. Grab a cow's tail, don't dig your toes in, and enjoy the ride.

Audra Brown has sadly not skied behind a cow.

CLEANOUTS SMELL LESS IN THE WINTER

There are a lot of jobs that are just not fun to do in the winter cold. There are a lot of jobs that are just about impossible to do in the freezing weather.

There are also jobs that are just the thing that I want to be doing in the months between spring and fall.

High on that list is the category of duties that one might truthfully refer to as stinky and always begin with the word clean.

Cleaning out a water tank definitely fits the criteria. Such projects vary in diameter from a yard across to twenty or thirty scoop-shovel widths, vary in depth from rubber soles to knee-high waders, vary in contents from caked sand to squirmy-monster flavored ice-cream, and vary in odor from two-week-old leftovers to pass-me-that-cow-patty-perfume.

The dry, non-pungent end of the spectrum is better in the winter because, in my opinion, anything that involves a shovel is better when it's not too hot. But on the other foot, the goopy, ghastly, odoriferous type of tank-cleaning which is, of course, the more populated part of the spectrum, is never fun, but I'm gonna go out on a board and say it's better to shovel frozen muck than thawed slime.

Such out-in-the-elements work does require the more moderate days of winter, but never fear, there is another job for those coldest of days. Tanks are outside the purview of heavy equipment, and therefore are de-yucked by hand and scoop.

Corrals, in contrast, have gates that tanks do not.

Hopefully, the front-end loader has a heater as well.

The process is much the same, just being done with a bigger scoop. Maybe with a skidsteer or a tractor, all the way up to a something that comes in a shade of construction-equipment yellow. Maybe all three in logical sequence to do the tidiest of jobs. Anything with an engine and bucket will do.

Don't quit 'til the diesel gels.

All that hay, mixed with manure, mixed with are-you-sure-that-used-to-be-sand, topped with balls of netwrap, evil twine, and even a bone or two, needs to be removed.

Audra Brown is articulate in the art of steering by articulation.

CATTLE CAN WAIT, UNTIL YOU WISH THEY HADN'T

Winter approaches quickly. Hopefully, all the peanuts and milo have been harvested, the hay cut and maybe even baled, the wheat planted and grown to a good stand, all the farming ready for what one hopes will be a wet winter.

As the daily demands of farming slow slightly from the normally busy rhythm, it becomes time to catch up on all the cattle-work that has been put-off and delayed in favor of some sort of seemingly more pressing business.

For anyone who owns cows, the matter of weaning calves needs to be addressed, if it hasn't already. Gather the herd and sort off all the calves but the handful that were born late and out of season. Give them their second round of vaccinations, something to keep the bugs off their back, and possibly something stronger to ward off any sickness that might try to take hold during this, their moment of stress.

As far as workings go, weaning isn't usually the worst. No branding, tagging, or gender-identifying this time around. That should have already been done. If it got put off all summer, then I'll shut my mouth. Weaning and branding all together is quite another bucket of oysters.

It's just like putting the two events together, but only needing to gather once.

That and the fact that five-hundred-pounders aren't quite as easy to manhandle and cut.

But either way, the result is similar. The cows get turned back out, but the calves stay in. They get fed the expensive feed that helps them cope and the freshest hay for a week or two. Then they usually get a free ride on a truck. And then it gets even better for these newly emancipated cattle.

Out they go onto the wheat patch for the rest of the winter with nothing to do but eat, sleep, and get fatter.

Or so one would wish, but the carefree cattle have a work-causing tendency to go and get sick and otherwise cause trouble. And so in contrast to his cattles' green pasture, the cattleman spends his winter days checking up on every single one of those calves. Tending their needs, putting out hay, fixing the water, practicing medicine, mending the fences, and as always, gathering strays.

Audra Brown says: wheat cattle are like British cars. The only day better than the one you get 'em, is the day you put 'em on that outgoing truck.

CRASHING THROUGH
THE SNOW

The snow falls, the water freezes, and there's lots of ice to break. But the cold can bring more than just work, and in between water tanks, it's easy to run into a good time.

Truth is, ground covered in snow (or mostly covered, it doesn't take as much as you might think) is considerably more fun than ground not covered in snow. Especially if you happen to have a semi-slick piece of something (such as an old pickup bedliner), a rope, some fellow farm-kids, and something to pull it all through the snow.

Put the sled behind, with the rope in between, and everyone pile on the pseudo-sleigh. Over the whitened wheatfields and through the muddy roads, hours of entertainment lie ahead with only the occasional pause to switch positions, re-orient the sled, or get a little work done.

The particular devices put together to create such fun vary, but the basic idea remains unchanged.
Sliding is fun.

Even when there isn't any snow, there is still sliding to be done. Out among the brush-covered sandhills, there can be found the occasional bare-faced mound. Those snow discs are great for sliding down.

Or, for the more experienced sand-slider, a cheap snowboard makes things a lot more exciting--if you don't fall down.

Sand isn't as cold as snow or ice, but it's a little more lingering when it gets in your clothes in a crash. But there is one thing far worse, as far as cleanliness is concerned; the true demonstration of how far a farm-kid will go, when there isn't a sandhill, and there isn't any snow.

It's big and tall, and nice and steep on the sides, and can be found quite often near fields and recently cleaned corrals. It's also green.

Snowboarding down the sides of a manure pile. Yes, it's smelly, but it's more fun than most of the other ways you get dirty on a farm.

Fun is where you find it--even if it smells.

Audra Brown always keeps a sacrificial snowboard in the back of the pickup.

FUN IS WHERE THE HAY IS

Hey out there.

Err...hay. Hay is out for the cattle. Right? It better be. I for one, don't want to have to do it in the snow. The heater in the loader tractor isn't much better than the air-conditioner... But it is better than the other place, there's nothing to use but a skid-loader up there and I still haven't thawed out from the last time I had to run an open cab in a blizzard.

Hay.

Snow.

Cattle.

Blizzards make me think of hay.

There's a lot of feeding hay that has to happen when the ground goes all white. Hay also makes for a pretty good shelter if it comes down to it. Stack it up around the outside of the corral if you get caught working cattle in the winter, or if your cattle are a little too short-haired.

Hay is also a lot like a playground. You can climb up and down and run around on top of it. I'm partial to long rows of round bales. Run up and down 'em (counting, of course, always need to know how many...and we definitely want to count 'em a few times.), jump across from row to row, and for the finale, slide down the side. Repeat until winded, rest a little, and go again.

I mean, shy of monkey bars, the party is pretty much covered. If there's some corrals with overheads over the gates nearby, or a certain sort of pipe rack, then it's definitely a playground.

But. I digress, as I am inclined to do. It's too cold to swing on the pipe rack and we were talking about hay. Which, of course, can be used to climb up (on a particularly tall pipe rack or some-such) by stacking and arranging it into what usually looks like a staircase for something with legs a lot longer than mine.

But on the ground, in the snow, in a pinch, one of my favorite winter usages of a hay bale is street sweeping.

I ain't kidding. I've seen it. Tractor with a hay bale, sweeping the snow off the highway, right through the middle of the town.

Audra Brown already fed hay.

NOT TOO WET, NOT TOO DRY

The tall fields of haygrazer are laid low, and as soon as the moisture gets just right, the nights of the baler begin. Many jobs can last long into the evening and the early morning, but most of them at least start in the daylight hours. Baling hay, in my experience, usually starts just after supper and ends at a time approximating breakfast.

The thing about hay is that after you cut it, is has to dry, but then, after it dries, it needs to be a little damp to bale.

Too dry and the baling process grinds the leaves into tiny pieces that is both less tasty to livestock, and less efficient. Too much of the hay never makes it into the bale and even more is lost when it is put out as feed.

Not to mention the fire hazard. Dry hay running around in a complex machine...all it takes is the idea of a spark.

The other side is even worse.

Dry hay keeps. It's maybe not the best hay, as discussed, but what gets in the bale is usually still around when you get ready to feed it. Wet hay doesn't get ground up and miss the bale. In fact, it shoves the loveliest, tightest, slickest bales you'd ever want out the back of the baler. But inside, the rot begins. Moldy, mildewy, maggoty feed isn't what you want to feed your livestock, they might not even eat it. But that's far from the worst of it.

Rotting organic matter produces heat.

Ever seen a big stack of hay burning? It probably wasn't a misplaced cigarette butt that started it. It was probably baled too wet.

Stacking a bunch of wet hay together is like building a giant timebomb. Each decomposing bale adds its heat to the others and with them all together, there's no ventilation or place for that heat to go. Eventually, it gets really hot, and then?

Spontaneous combustion.

Suddenly, the core of the stack is burning. It burns outwards and before you know it, you have a haystack fire. And you can't fight a hay fire. You can keep it from spreading, you can stir around the burning bales to get it over with quicker, but there's no saving the stack.

Don't bale your hay too wet.

Audra Brown says: Small stacks save bales.

SAME PLACE, SAME SONG, SAME TIME

There are certain jobs on the farm that require long stretches of seat-time in order to get it done. Any harvest, planting, cultivating, or maintaining operation usually amounts to a sequence of contiguous days and or nights running the same piece of equipment, doing the same sort of thing, until it is done. These repetitive jobs are generally repeated on some sort of regular basis.

Take wheat harvest, it happens about the same time in late-June and early-July every year (or hopefully every year). It involves the same pieces of equipment, and usually the same people do the same job every year. Big sister on the combine, little brother on the grain cart, and baby sister at-large.

So, same job, same locations, same machine, same crop, same time.

The years and the circles tend to run together.

But some songs seem to stick.

That essential piece of any tractor or combine, the AM/FM radio, helps to keep the monotony at bay. However, if you listen to the same station or two (or even three) all day long every day for any length of time, you start hearing the same songs and patterns start to develop.

In my experience, most harvests end up with 2-4 new songs attached. You'll hear a new song you like for the first time during that harvest and then you'll hear it over and over again for days on end. Sometimes, in the midst of all those equally round fields and equally waving golden wheat, you'll remember a particular pass through a particular field when that particular song was heard for the first time.

Years later, you might still be able to recall the set of four or so songs that defined that harvest musically, and along with it, the particular breakdown that may have happened or maybe the odd, but edible lunch that finally showed up a little too late in the afternoon.

And every time you hear one of those songs on the radio now, you automatically think of the other three songs and the field of wheat where you heard 'em and so on and so forth and you'll probably never forget, even after you can't remember your own phone number.

Audra Brown forgot what song is stuck in her head.

THE FINE ART OF LOADING

Let's tie down!

Words that you may or may not be pleased to hear.

Words that are heard quite often on, around, and on the way to, the farm and/or the ranch.

Perhaps the most common, is when it's time to secure a relatively simple and homogenous load. Hay bales, pallets of feed, or sacks of seed.

In such cases, strapping down a load involves repetition and quantity. Many straps with a couple of rounds of tightening and all in rather linear formation. Straight up and over the items, repeat until everything is cinched down. Easy.

Square bales load quite neatly. Beyond the obvious issue, round bales are straightforward enough. Pallets of fifty-pound bags ride well if a properly excessive amount of shrink wrap is applied...otherwise, they have a tendency to slide around and fall off and there's never enough straps.

But not all loads are so nice and neat.

Trailers get loaded with more stuff than they should and things that are never intended to be loaded.

In chaos-transport situations such as auctions, auctions, auctions, and the occasional transport expedition to another piece of property, the tying down is secondary to the particular puzzle that is the loading itself.

It's akin to a deadly combination of three-dimensional Tetris and Jenga, where the pieces are not intended to fit together, everything is heavy, sharp, and you and/or your fingers are never far away from the descending blocks.

Then there is the consideration of fragility.

There are always things that are so physically awkward to load that it seems impossible until at least a couple of minutes after it is successfully loaded.

Greatest hits include, but are far from limited to, loading three bobtail trucks, grain-beds and all, onto a couple of semi drop-deck flatbeds. The first two weren't so bad, one big forklift and a hefty skid-loader on one side, and a good-sized CAT loader on the other and you can lift rather large things up into the air. A bobtail with its bed loaded with dirt, is not one of those things. For that, only one end at a time can be lifted, and just barely.

But it can be done.

Audra Brown treats every trailer like its loaded.

THE GREAT GATE THEORY

There are theories and equations that are undecided and constantly debated and improved. A lot of thought and time and money is spent analyzing the motions and patterns of the universe.

One theory in particular is of great concern out on the farm or on the ranch.

The optimization of a route for the minimization of gate-getting...

....while still getting to the destination in time to get any work done.

This conundrum is deadly serious and infinitely more important than such things as how to make a profit, how to to get more work done, and how not to get stuck in a pile of tumbleweeds.

The parameters of interest to this theory are the number of gates, the types of gates, and whether the gates need to be immediately closed or could wait until the return trip.

You might (or might not) be surprised at the lengths to which any sensible person will go--or go around, in order to reduce the number of gates to be opened, shut, or speak-of-it-not...both.

A mile out of the way, for example, is no decision at all.

Five miles might be noted.

Ten miles could go either way.

If ever you observe a movement of a group of farm and ranch vehicles in their natural habitat of gate-infested land, there is a notable and serious maneuvering for position prior to being locked into caravan order by the narrow, one-wide roads.

The vehicle that leaves first is always strange. Either open-cabbed and trading less dusty air for a gate, or too impatient to wait. As soon as that odd soul starts the migration, all the remaining vehicles rush to get in line and not, for heaven's sakes, have to bring up the rear.

It seems a small thing, to step out and open the gate, and just as easy to close it in much the same way. But it is not. Gates left open have led to plenty of livestock having misplaced themselves and few lost strangers getting themselves stuck in the sand.

Boats use nauts, cars use miles, overseas they use kilometers, but distance is better measured in gates-you-have-to-get.

Audra Brown gets gates when there is no other choice.

GATES OF FIRE AND ICE

In the constant quest to keep cattle (and other livestock) contained, the gate is a crucial component.

In the best of times, the great swinging gate holds the line. In some places, the gates of doom hang over our head with the threat of well-weighted assistance, and elsewhere, the gates of air and aluminum swing freely in the breeze.

But here in Realsville, not all gates swing so well.

Many will never swing. Some will never truly be gates. Some are certainly not gates at all.

Gates rigged out of pallets, panels, and pieces of equipment are always around. We call them gates for lack of a better term usable in pleasant company.

The gates of fire, are traditionally seasonal warriors, showing up during the winter months when cattle are quartered on fields of wheat or other temporary pasture that does not have a permanent fence. These gates are not much to look at, appearing as little more than a wire strung across the road. Nonetheless, they are charged with enough electricity to make contact quite unpleasant. They are required to wear flags of bright color or duct-tape silver in order to be more visible to both their prisoners and their allies.

These unassuming gates and their accompanying fence have one glaring weakness.

Tumbleweeds.

These rolling menaces of the plains are quite good at breaking and scattering electric fence.

Luckily, preparations can be made to ward the wire against the great tumbling menace.

By constructing the fence and gate of two-ply wire, either barbed or not, many tumbleweeds will be stopped in their tracks rather than bust through and break the crucial connection to the fencer.

The other enemy of these gates is, sadly, their allies.

Due to the come and go nature of the gates of fire and their Spartan appearance, many fall to the vehicles of their builders when they are forgotten, or for whatever reason, go unseen.

They are always repaired, but as these valiant gates mature, they collect many splices and knots until the day when they can no longer hold themselves together and are reduced to so many pieces of tie-wire.

Audra Brown has had to splice a few gates.

BEWARE THE GATES OF DOOM

Getting gates is no fun, running gates is a lot of work, and doing either in the wind is likely to cause bodily harm to something. Nonetheless, gates are very important and the kind of gate can make all the difference.

The most well-balanced and always appropriate sort of gate, is a swinging gate.

Constructed of otherwise unimpressive pipe, wide enough to fill the gap in the fence without fitting too tight, tall enough to keep all but the most air-inclined livestock in (or out), bars solid and of a distance apart as to make climbing it in a hurry as painless as possible, a diagonal reinforcement to help with the sag, a quick, secure, and easy to understand handle/latch, and one set of well-fitted, well-oiled hinges.

That's not too much to ask, right?

Seems it is.

Gates of such simple, yet optimal function are by far the minority.

There are those swinging gates that might at first appear to be such as described above, but at closer inspection, sometimes from beneath, are not.

They are so heavy that neither the hinges, nor you, can handle the weight.

If the hinges do give up the fight, don't stand there. Move.

Lighter would be better than that, one would think.

Then one would also need to decide whether it is better to have a gate or a two-thousand pound bull in your lap.

I'm presently undecided.

I have seen the backs broken by massive gates of doom and also the crumpled, tangled mess of aluminum and/or light pipe that had mistakenly masqueraded as an obstacle to large, moving masses of future steak.

Gates are like a lot of things. The amount of thought, time, work, and money you put in them, will be well worth it for decades down the road.

Good gates are always worth it.

Bad gates are always trouble.

Audra Brown has put a ring of chain around the gates of doom.

THE GATE: UNCHAINED

Don't spit into the wind.

Good advice in this part of the world. We all know somebody who spits, and we dang sure know about the wind. There's a lot of good advice out there. Here's the trite wisdom for the day, brought to you by our proud sponsor, experience.

Chain the gate.

That's it. Three words that will save you more trouble than you can count, even if you can count quite high. Ignore this advice at your own risk. It applies to every possible combination of gate and chain--such that it is physically possible. In all other cases, adjust the definition of gate and/or chain until it is possible.

Every rule is based on an incident, and the best rules are based on the worst incidents. This rule is no exception. Allow me to convince you of its legitimacy.

In the working of cattle, gates are a common and irreplaceable installation. They could also deserve one of those "death or serious injury" labels. Some gates have handles and other quick, convenient latches to hold them closed. Don't be fooled. Chain 'em.

Unchained gates open at the most inopportune times.

Such as, when you're not looking.

Getting hit in the back and slammed into the ground or into the fence, or drug underneath, or into the path of a charging bovine, or...you get the point. You don't want to get hit by a gate. So, chain it.

But you say you don't care about your own safety or the lives of others? How 'bout money? Nobody wants to see money disappear. If the gate separating your calves from the outside world gets open and they depart for parts unknown, there goes your dough. At recent prices, that's a loss of like 6 months of gas in your pickup for every four-feet that disappears. That's a lot of dollars at ten mpg and five miles to the mailbox. Thus, chain the gate.

Now for the nightmare.

It's the story that little keeps little farm and ranch kids awake at night.

The gate on the trailer: Unchained.

Imagine dropping a large animal on the highway at 65mph. The results are never okay. Backs are broken, hide is scraped, heads ache, hearts break, and some small imports are never identified.

Audra Brown doesn't want to see the epitaph,
"She didn't chain the gate."

THE ROAD ISN'T PAVED WITH INTENTIONS OR ANYTHING ELSE

Work boots and ground clearance. You need one to get there and the other to get anything done.

Friends from town used to come out and see me and I usually had to pick them up at the gate. Work boots weren't even an option, the best you could hope for was that they had some sneakers that they didn't mind getting dirty.

They got to ride on the tractor with me, or whatever equipment I was running that day. I know it wasn't what they expected, but it was as good as I could do.

A day off to run around and do nothing wasn't in the cards, but a game or two of go-fish wasn't out of the question, as long as I didn't have to deal, hold the draw pile, or look at my cards during a turn.

To me, it was great fun. Talking back to talk radio, singing alone or with the radio on, reading with one eye, or even trying to talk on the phone was no substitute for live conversation and the laughing and joking that friends can do. Maybe it wasn't quite so wonderful if your perspective was more urban, but I hope they enjoyed it as the glimpse into my life that it was.

It was rarely an occurrence and almost never a repeat event, but I can't really blame anyone for staying with what they know. I surely felt similarly when I was the one sleeping over in town.

I never knew what to do when there was no work to be done.

When you grow up on the farm, and the gap between the driveway and the highway is a few miles of sand, your first car is a pickup, and if you're me, the bigger the vehicle, the more comfortable and easy it is to drive.

And that one time you drove your grandparents' town car home from church?

That was about as nerve-wracking as it gets.

Looking in my closet, there's more worn out work boots than all the tennis shoes I've ever owned. The motorized vehicle I've owned the longest is tall, dirty, and drinks gas like I guzzle iced tea. These days I do like to wear sandals, and my stable even includes some cars, but as much fun as it is to drive a go-kart, inclement weather means a stop somewhere shy of my driveway and a switchover to my good ol' pickup that will make it all the way home.

Audra Brown learned to drive a long time before she learned to drive a car.

MESQUITE ON A MOTORCYCLE

Beware the badness of the terrace...

...and the terror that sometimes grows on it.

Farming and ranching takes a lot of country in the desert. Miles of pasture or fields. Lots of ground, and lots of variety. Thus, it would behoove one to invest in something that is deemed to be "all-terrain."

Alas, such a thing is still little more than a fable and an over-used advertising scheme. There are certainly, "off-road" and "multi-terrain" contraptions, but I have yet to find one that would actually propel itself successfully over all terrain.

A big bulldozer is the most unstoppable rig I've found so far, but it is just not practical for a great many things that need doing. Indeed, speed, maneuverability, and the fact that if you drive it across a proper road, the road is then in need of repair, makes it unusable in most of the common situations where "all-terrain" is what I need.

And boy did I encounter a little terrain the other day.

There were cattle to be gathered and both a surplus of area and terrain variation where those bovines were located.

66

Starting on the fields of milo stalks they were grazing wasn't bad. The manure piles were the first interlude of terrain that was less than passable. I really wasn't in the mood to bury my cycle in a molten methane trap. Luckily, no hapless bovines had fallen prey either.

There were a few mud-holes and some rough grassland pasture to cross to get the cattle into the corrals, but mishaps were pleasantly minimal and before too long, the cows were in the pen.

All but two.

Of course. All but two. And where were they but in the thickest, densest, most miserable bit of terrain for miles. If you've never chased cows through a mesquite patch on a motorcycle, be smart and keep it that way. My tired were flattish from all the pinholes from mesquite encounters of the past, but I still cringed every time I propped my feet up on the handlebars and reluctantly drove through a thicket that could not be avoided. As those two were not highly cooperative, high speed was frequently required. No easing the pain of passage through those thorns.

It was hard on clothes, hard on tires, and hard to enjoy, but it had to be done. Now that I think about it, maybe I need a vehicle that can handle "less-terrain."

Audra Brown knows better, but one can imagine.

BIG BULL, SMALL PEN

Gathering cattle is a job that can vary in scale from a small corral to a small country.

Differing conveyances and tactics can be used to accomplish the desired wrangling of bovines, but it might be surprising how many different ways cows can be punched.

In the corral, where space is tight and cattle frequently have nowhere to go but over (either you or the fence), on foot is popular due to the maneuverability, visibility, and well, mostly the maneuvering. The cattle are controlled primarily via psychological manipulation, and maybe a stick or whip to direct attention. Fences should always be climbable, and the nearest climbable fence should always be in one's mind. Just in case.

The next step up in the corral situation is onto a horse. Giving up the subtle and infinitely controllable mobility of afoot, one gains mass equalization or better, while retaining a very effective level of maneuverability. (It also means not walking, or running, and that's always a plus.) The few places in a corral where a horse cannot go, are probably hoof-cannon firing lanes and should be avoided anyway.

Then, there are bulls, of the stubborn variety---or worse, the anything-other-than-calm variety. Such large and muscular creatures are too big and quick to be around if they are even the slightest register above mellow.

In such cases, the work must get done and the bull must be persuaded to move to the desired location. You can't wrestle on foot, you can sometimes wrestle on a horse, but you can always wrestle in a loader (or other heavy piece of equipment.)

If you don't have a loader around when you are working cattle in a pen, you must not have a loader. They are not always needed, but between feeding hay, rescuing bovines in unlikely positions, and encouraging immovable beef to move, they are irreplaceable. Cattle may not understand mass and physics, but they can sense weakness. If they are disinclined to cooperate and are being encouraged by a horse, or flimsy gate, or anything that gives an inch, they will fight on.

And on and on.

Hopefully, the location of the stubbornly stuck steer (or other bovine) allows for the insertion of a loader. The bovine will quickly stop being so stubborn once it senses the superior immovability of the loader. Otherwise, one might have to get creative. A situation involving a ratcheting strap and a hefty Hereford comes to mind.

Audra Brown knows where the fence is.

KEEP YOUR PRIORITIES STRAIGHT

There are priorities--and then there are Priorities. One needs to be done before other things get done. The latter probably didn't exist five minutes ago and it has to be done now. No waiting, skip to the front of the line, do not pass go, do not go check those two-hundred heifers. In agriculture, there are plenty of both, and honestly, too many to ever get done.

Some priorities might be better described as chronic priorities. Once you contract them, they never leave. The best you can do is take care of them as often as possible and they might recede for a few hours or a couple of days.

Keeping cattle fed, watered, and in the intended pasture. Keeping fence up, unburied, and moderately capable of keeping adventurous livestock on the proper side. Keeping the sprinklers moving and full of water. Keeping fuel in your pickup.

Chronic priorities are always important, but due to their continuous nature, all but the most time-sensitive, frequently are forced to take a back-seat to more pressing matters. Checking the cows water might be pushed back a few hours or even a day. Fixing the water once you know they are out takes a back-seat to nothing except keeping fuel in the tank and tools in the hand.

Known priorities are usually more cut and dried than chronic ones. They are the expected tasks, the scheduled tasks, the steps in a process that get crops going, barns built, and cattle branded. Once begun, they take the top spot until they are done, but they usually have a little leeway with regards to which week is dedicated to plowing, planting, branding, or building.

Prime priorities are frequently a product or piece of those big, extended priorities that goes wrong or comes up missing sometime after you start and eliminates the possibility of timely or proper completion with its absence.

Say for example, being 70 doses of bovine, modified-live vaccine short, one-hundred miles from the nearest vet supply...on the later half of Saturday. Half the cattle have already had their shots and some having an infectious vaccine in the same pen with some who have not, over the whole weekend, is not a recipe for success.

When the one person who is not at the cattle-working gets that call, homework, food, plans...they all go out the window and vaccine acquisition and delivery is all that matters.

Audra Brown arrived two calves before they ran out of vaccine.

EMERGENCY RATIONS

Now, there are some mighty tasty vittles in the world...

...and there are things that are edible in the technical sense only.

Consider F-250s and El Caminos. Alike only in the broadest of technical definitions.

Similarly, there is a wide range of substances that will sustain-life, some, more endurable than others.

On the less-appetizing end of the spectrum, there are the staples that are equally awful and wonderful at the same time. The things that live under the seat. The cans that will save your life (or at least it'll seem that way.)

I refer to them as emergency rations.

They live under the seat, in the toolbox, and everywhere inaccessible in-between. Cans with self-opening lids. Sardines in Gourmet Mustard Sauce and Vienna Sausages of all slimy varieties are my personal recommendations.

Emergency rations are a sort of paradox. If they are in the least bit appetizing from day to day, they wont last long enough to be there when you need them most.

They have to be a little disgusting,
and more than a little slimy.

They have to be so unappetizing that you wouldn't even
think of snacking on them until that day arrives when
you've been at it all day, nobody is gonna bring you
anything to eat, it's been hours since you should have
eaten, you've still got hours before you can even think
about leaving the job you are working on...and if you
don't eat something soon, you swear you're just gonna
fall down and die.

It might not be strictly true, but inside the moment, it
seems even worse. And that's when you reach under the
seat and pull out a dust-covered can of pure happiness,
suspended in a heavenly slime.

Pop off the lid and start scooping out those strangely-
colored morsels of probably-meat. They slide right down
and if you didn't know better, they taste better than a
convenience-store burrito.

You wash it all down with the au jus, of course, and then
you go back to work.

Audra Brown prefers classic viennies.

TRUCK TURNAROUNDS

Where a vehicle goes is an important part of what a vehicle is and frequently, a part of what it is called.

It is important to know where a vehicle goes.

True off-road rigs like a steel-tracked bulldozer, can go practically anywhere, but that doesn't mean they should. If you go clanking across a road, or down a road, the road will suffer in direct proportion to how nice it was. The better the road, the higher the technology, the more damage is done.

Dirt recovers, caliche can take it, but asphalt is a mess.

The not-quite true, but wishfully labeled vehicles called all-terrain, are in the middle. Capable of many non-road applications, they are less likely to treat roads as disposable, one-time-use infrastructure.

Then there are those delicate automobiles that are on-road only.

Cars come to mind, but they honestly have little reason to leave a road, and less reason to do so on a farm. At best we might have one to go to town in, or one in the barn to work on during the long winter nights.

But there is a beast far more persnickety and far more useful that is truly made for only the better developed sorts of roads. I call it a truck, but to be specific, let us refer to it as a semi.

Hundreds of thousands of pounds of cattle, grain, hay, and other harvests have to be hauled off to town to be sold.

In short, they must be trucked.

There is no other vehicle for the job, but it was never designed to drive through the sand, the mud, or the uneven pasture. But with a little luck and a lot of momentum, it can get to the load. And if you can get it in, you can usually get it out.

But the true test is getting a truck turned around.

In the corners of fields, near corrals, and headquarters, large loops of abnormally good road get constructed. These permanent bits of infrastructure have only one purpose and get used only a few times a year, but they are truck-turnarounds.

A noun, a landmark, and a rallying point, now you know what to think, the next time you hear me refer to a truck-turnaround.

Audra Brown has turned a truck or two around.

VARMINTS, NOT TO BE CONFUSED WITH CRITTERS

Plowing, planting, branding, weaning, harvesting, herding. Agriculture keeps you plenty busy, but you might not believe how much time is spent on pest-control. Fact is, there's never enough time for this particular duty.

It's almost as never-ending as building fence.

On the lower end of the size-spectrum and the upper end of the priority list, are the bugs.

Insects and worms, larval and mature; bite-sized varmints are responsible for many a lost crop. One uncontrolled hatch of army-worms can turn a fine, green, growing wheat field into a dead, worthless patch of dirt. And the only thing that will do a thing about it is a timely application of pesticide.

But there are other, bigger, not so easy to handle pests that plague us. I tend to call 'em varmints.

Now, there is some confusion as to the terminology of what constitutes a varmint as opposed to say, a critter. In my lexicon, a varmint is any sort of critter that consistently causes harm to crops or livestock.

Examples of varmints include: coyotes, rattlesnakes, mountain lions, prairie dogs, feral swine, certain avian species, and anything else that might do damage (porcupines will have to get their own write-up, not enough time today.)

Predators like coyotes and lions will kill cattle if you let 'em and one dead calf is like somebody eating a month of paychecks.

Rattlers are the same, if deadly in a different manner.

PDs are little evil lawnmowers that cut down any vegetation for several yards around each hole until there is nothing left alive. They kill grass, crops, everything around them. They can turn a field or a pasture into useless dirt almost as fast as a wildfire--but at least with a fire, it'll grow back.

Feral swine are not always a problem in these parts, but when they get a foothold, they are a terrible pest. They are not only dangerous to people and animals, they will root around and dig up crops.

There are other varmints around and the one thing they have in common is that getting rid of one makes us and our food safer.

Audra Brown knows one when she sees one.

CRITTERS,
NOT AS CUTE AS THEY SEEM

Last week, we discussed the definition of "varmint." This time, I'd like to talk about critters.

Now, critters are not, by definition, excluded from being varmints from time to time, but they're generally not full-time varmints, or might never really be quite as bad as a varmint. Also, a lot of critters make good eating.

Of course, despite a general lack of extreme damage-causing, neither are critters particularly beneficial. They can be very big problems in small packages.

Some commonly known critters include: jackrabbits, cottontails, bull snakes, hog-nose snakes, racer snakes, king snakes, deer, antelope, and most birds.

Some of these have beneficial uses. Non-venomous snakes are good at keeping down the rodent population, deer and antelope make good jerky, quail tastes better than chicken, and rabbits...well, rabbits make stew if you're hungry. Rabbit meat isn't one of my personal favorites.

But any of these critters can become worthy of the varmint categorization in certain circumstances.

The easiest and most frequent is overpopulation.

In anything but an amazingly lush year, too many herbivores can seriously cut into the feed supply for cattle or other livestock.

Too many snakes just makes things a little too edgy. Constantly wondering if it's a rattler or a mundane snake is hard on the nerves.

And birds eat too, easily turning into unwanted crop stealers if they get very thick.

I'd say disease is the next thing that can easily turn a relatively mundane critter into a problem. Birds, especially, can be a real issue with livestock. They get everywhere--let's be honest, they poop everywhere and on everything, spreading bacteria and disease as they go. In small amounts, it's unnoticeable and irrelevant. But just a hair too much, especially with young, stressed livestock with already overworked immune systems, and you've got monstrous problems.

Not a few lives and banks have been broken when the wrong strain of the wrong illness gets hold of a cattle herd.

Point is, critters can be cute, but don't underestimate the little rascals, and don't be afraid to do what needs to be done.

Audra Brown is neither a varmint, nor a critter.

HAVE YOUR CAKE AND FEED IT TOO

Once again, it's time to delve into the lexicon of the agriculture business.

Today's word is cake.

It is a noun, a verb, an adjective, and has nothing to do with sweet, spongy dessert or what Lucifer likes to eat. An example of its use as a noun might be somewhat confusing if overheard out of context. "I'm gonna be gettin' a semi-load of cake, next week."

The human-preferred cake is usually spoken of in terms of baking, taking, eating, or making. A good guideline for making sense of conversations involving cake is to pay attention to the size of the measurements.

People cake is not often spoken of in quantities such as truck-loads, bin-fulls, sacks, pallets, tons, pounds, feeders, dumps, or cubes. These measurements are more typical of the bovine equivalent. Study up, there may be word problems later.

If Old MacDonald had 13 cows and the middle hopper on his thirty-ton overhead bin is about half full, how many pallets of cake does he need to feed three pounds twice a week for the rest of the winter?

Let's see...40 sacks to the pallet, three pounds to the dump, four legs to a cow, three seats in the feed-pickup, and nothing in the bank. A truck-load, final answer. Bulk too, no time to feed 'em one sack at a time.

A truck-load is always the answer...or I may be a few cubes shy of a bin-full.

Cattle cake, often called cattle cubes, is surprisingly, not very cubish. Tubular pellets of compacted vegetarian protein, of lengths up to an impressive seven or eight inches, but usually checking in around three or four. A dull, medium shade of variegated brown, they don't look like much, but they taste better than they appear. Popular diameters are three-quarters and half-inch, with my own experience suggesting a that the former is preferred.

Don't forget to cake the cows.
May your feeder never run dry.
May your count always be right the first time.

And the next time you hear the wrong side of a phone call between a rancher and his cake-dealer. You'll better understand: "Yeah, I need another truck-load of three-quarter, 20% cubes delivered."

Audra Brown would like dessert better if you could specify a high-protein percentage.

SOME HORSES HAVE HOOVES, SOME HAVE TIRES

A horse is a horse, of course, of course---but sometimes it has more wheels than hooves.

Now, I love a good equine-driven cattle gathering, I love riding, but the thing about horses is that you have to catch 'em. On a good day, they're at the water, bored, and ready to come see what new adventure or new something to chew on you might have brought. On a bad day, you end up afoot, walking across the pasture, and in circles, trying to get a halter on a horse.

Then, you've got get 'em back to where you left the saddle, brush the stickers off, and get all the gear on and cinched up just right. Then, you've got to get the horse to the cattle, which, depending on how far that is, can be a chore in and of itself. If they're pretty close, you could ride over, if it's a little further, you can just haul them in the trailer, and if it's out of your district, you've got to have hauling papers too.

However you get there, a horse is handy and good for the canyons and breaks where the terrain is extra rough and the cattle are too. But even on a fast horse, you don't just go from one place to another on a whim.

You're stuck and subject to the level of stubborn your horse is being.

And they eat a lot, all the time, even when you don't
need them.

These days, the four-legged horse only gets out on
special occasions and really big jobs. The rest of the
time (and since cattle are always getting out, that's pretty
often), the two-wheeled horse gets the job.

It only eats when it's working and it will get me there
faster than a pickup.

And it's never hard to catch.

In fact, it's usually just exactly where I left it the last time.

It does sometimes need a good kick to get started, but
after that it runs all day without complaint.

Don't tell anybody, but sometimes all I have to do
is show up. The regular bovine escapees who make
fence ignorance a habit can hear it coming and after
a few experiences, they know that sound means the
motorcycle has arrived to put them back where they
belong. They hear it, and with a resigned moo, turn
around and head back in.

Audra Brown has a hooved horse and a wheeled horse,
and they're both a bit tall.

A PAIR IS A PAIR,
EXCEPT WHEN IT ISN'T

Cattle operations all have specific accepted terminology. In what is commonly referred to as a Cow-Calf operation, we use the term, "pair," a lot.

This word, in common vernacular, is relatively nonspecific, usually meaning two of something, and often with the implied idea that the two somethings are alike or a matched set in some way.

In a Cow-Calf operation, a pair is a very specific thing. A pair is a cow and a calf, where that cow is the mother of that calf. Most of the time, the cow is the birth mother of the calf, but on occasion, adoptive pairs are formed. The term also has an associated verb-form, "pairing," which refers to the act of trying to figure out which cows go with which calves.

The vast majority of the time, this terminology is perfectly appropriate. But, just like any label, it does fail. Deeming a set of a cow and her calf as a pair, makes sense. It is a set of two, related/matched in some way, and unlike some species, cattle only take care of one offspring at a time.

Except for the rare---but always welcome---event of twins.

This does happen often enough that we're all aware of the possibility and so are not overly surprised when a cow pairs with two calves. It does make for some interesting stories.

Like just this last week, some cows, mostly with calves, needed to be moved from one place to another. They got gathered, sorted, put on a truck, and unloaded a few miles away. One of these cows had a very new baby with her. He was big enough to travel, but surely only a few days old. Business as usual.

One cow. One calf. A pair.

The next day, they were gone. We found them back on the other place. And guess what, they weren't really a pair. She'd gone back for another calf. Twins, we thought, pleasantly surprised, and we left her and her two calves where they were and made plans to come and get them when we had the time.

But, you know what happens so rarely as to equate with never? That's what happened. Next day, there she was, far from a pair. She had all her calves. We counted, one, two and wait for it...three!

Audra Brown has seen two sets of triplets this year, doubling her prior life count.

KOW-RATE:
THE ART OF PUNCHING AND
KICKING COWS

There are many martial arts, some new, some old, and some in-between. But I'd like to talk about one of the oldest, yet never-talked about martial arts. The delicate art of Kow-rate.

It's origins are shrouded in green, methane-scented, mystery, but it's practice is something I can tell you a little about.

I have personally been practicing the art since a very young age.

The first techniques learned are generally defensive or herding moves. Like the use of a cattle-prod, whip, or one's own body to move cattle in particular directions or patterns; also, the ever important fence-climbing defense. A critical move that can later be developed into the advanced techniques of fence-hurdling, or even fence-jumping, the discipline of the Longhorn.

The later disciplines are not necessarily learned in any order, and one may choose to pursue a few to mastery and let others concern themselves with the other disciplines. I am particularly versed in the art of the Head, in the area of the Branding.

It might be compared to the human-directed arts of wrestling or maybe jujitsu. The way of the head is very intimate and very physical.

A practitioner learns to move the hands faster than the ears of a bovine; becomes familiar with all the features of the head, veins, pressure points, horns, etc...; and finally learns to control the bovine's powerful neck in various ways.

In the chute, the one-armed nose-hold is quite challenging and the catch-and-throw for when the chute isn't going to work can get very exciting, but the way of the Head is a good way.

And it's the most visually notable.

Those who practice the Head, can be spotted after a branding by the red splatter all over their clothes and face and hat.

Other aspects of Branding include the way of the Hot Iron, the path of the Dull Blade, the Fist of Sharps, and the Ones Who Work Behind.

There are many more areas of this old and storied realm of study. But not for today

Audra Brown is punching something.

THE ONES WHO WORK BEHIND

Let us once again consider the ancient martial art of Kow-rate.

In the area of branding, we have already discussed the Way of the Head, or the Red Way. A very hands-on discipline, it is marked by nose locks, ear grabs, and proficiency in a variety of tools.

Today, we shall consider the mastery of the other end. Those who pursue the disciple involved in getting the cattle to the chute in the first place. Practitioners of the Green Way.

The Ones Who Work Behind.

The Green way is a job with little thanks and less grandeur. It is dusty, dangerous, and frequently lonesome. While others work together around the chute, capable, if not always inclined to converse, the one in the back can only holler, and since that is part of their job, rarely gets any attention.

Armed with a cattle whip, a poker, a hotshot, or only their presence, the ones who work behind move the cattle from pen to pen to holding-alley to working-alley and eventually up to and into the chute. More so than the others, they must enter the pens and walk among the cattle, and behind the cattle.

Hoof cannons and tail throws are very real threats that must be avoided. The ones who work behind must be quick and lithe to avoid these and other issues that come from being so close to large, moving bovines.

Those practicing the Green Way can be identified by the thick layer of dust that covers their being and the green marks of where bovine tail throws landed and left their pungent imprint. They are frequently seen wearing masks to protect their face from the dust and green, and are usually quite possessive of their tool of choice.

Never come between One Who Works Behind and his or her favorite whip or poker. For they are skilled and likely to use it on you. Their reach is long and their patience with people is less than their patience with cows. Treat them with great respect and care.

They are irreplaceable in the area of branding and their expertise crosses the areas more than any other. Give them a new bandanna or a whip every now and then because without the Ones Who Work Behind, nothing would ever get done again.

Audra Brown tosses an imaginary Coke to the ones who work behind.

A VIEW FROM THE ALLEY GATE

"Heifer!" The sorter at the far end of the alley yells with confidence.

As the hand running the gate of the pen which has been designated the sorted destination of such young female bovines, you scramble to open the gate as the heifer runs your way.

She trots right at you, at least one eye on your person, and at a point that seems very late and very close, she jumps sideways, maintaining her view of you as long as possible, before making her way unerringly through the hole you opened and into the pen where she belongs.

The Spring sorting has begun, only a couple hundred more cattle to go.

Shouts of "Steer!" let you close your gate and watch 800 pounds fly by.

"No!" or "Both!" or any grunt or growl of either confusion or exasperation means close all the gates and pay attention.

You may need to grab another gate, or jump out in front of a charging bovine to try and get them sorted.

Or, you might just need to levitate onto the nearest fence-like structure and wait for them to either settle down, knock themselves sensible, or go to the back of the bunch and delay being dealt with for a few short minutes.

Then comes the bunch that has to be sorted more than two ways.

Eartags of at least three different colors shouldn't be too bad, and it isn't for the most part, but for when the colors start to run together.

Blue and pink don't get mixed up.

Add yellow and green and it's still pretty clear.

But when orange and red and pink are all in the same bunch, mistakes are made and have to be made up.

Eventually the dust clears a little and everything is sorted just like we planned. Then we weigh 'em, feed em, and turn 'em back together again.

Audra Brown ran the gate, the scales, and the chuckwagon.

THE TRUTH ABOUT LONGHORNS

'Tis the season for bull-buying.

The choices are many and so are the dollars.

It's a critical decision point for any cattleman. For a beef producer like me, the end game it's all about pounds, and color. That's right. Color. Why? I'd love to know. But the truth is that black cattle will bring a premium that other shades don't.

And spots? Don't even get me started. Some cattle-buyers will cut out a spotted calf faster than a cripple. But I just hauled three, skinny little longhorn bulls home. Considering the downsides, why would a longhorn be important in a beef operation?

Two words.

Calving heifers.

Heifers are female bovines before they've had the first calf and are a special sort of trouble. Calving for the first time is an iffy business and a big, pretty, beefy baby doesn't help.

Enter the longhorn bull.

That's the bull we put on the heifers that first time around because it keeps those first calves small and makes calving as close to easy as it can get. Sure, you might get some spots, and they'll never weigh like you want 'em to, but the extra sleep I get at night is well worth it.

Longhorns have their faults. Skinny in the rear, pointy up in front, and a little too good at jumping over fences, but I've got a soft-spot for the breed.

Ease of calving, by far the main reason for using longhorn bulls to breed the heifers, isn't the only perk. Longhorns are smart and resilient. I've seen longhorns live on not much more than sand and water. Longhorn cows make the best mamas, taking care of her calf, fighting off varmints, and they live a long time.

My grandma had a longhorn cow. That old thing had a calf every year, took good care of it, and lived to be twice the age of most cows. The last time I saw her, she was on her way to the sale, and so old we'd all lost count.

She also had four horns.

I doubt I'll ever see a cow as cool as her again, but if I do, I'll be reaching for my checkbook.

Audra Brown wants a whole herd of four-horned cows.

TO FEED OR NOT TO FEED

Cattle come in a range of temperaments.

The spectrum runs from:
"I-can-see-you-therefore-I-will-run-over-you"
to
"I-shall-keep-my-nose-in-your-pocket-until-I-find-the-feed-that-I-know-you-have-in-there."

Either end of the spectrum can be dangerous.

I prefer mine somewhere in the middle. Wild enough that they properly respond to the stimulus of a human presence by moving away from it; gentle enough that they move away at a leisurely pace.

Alas, few cattle demonstrate such an optimally moderate behavior.

On the one end, cattle that we in the business refer to as, "crazy," and they come in two subtypes.
"Mean" and "Wild."

Mean and crazy cows move fast and in your direction. They treat every encounter with a human as the ultimate game of chicken--and they never lose.

They can be distinguished from their close-cousins, wild-crazy-cattle, in two ways.

In a pen, the mean ones don't need a reason to head your way. Sometimes, they don't even need a clear path. They are the ones that will bloody their noses by hitting a steel fence that a human is on the other side of, or knock the fence down on top of you if the fence is less than well-supported. They are that cow that you coax through the gate or into the trailer by using yourself as bait.

Remove the 'mean' and they run in high gear all the time, sometimes away, sometimes at you, but at least they will usually veer off at the last second rather than running right over the top of you.

Both are difficult. Neither are gentle.

Far across the spectrum, there are those cattle that are literally so friendly and gentle as to be dangerous and difficult. They don't fear humans--at all, and therefore, they don't herd because they don't feel the need to move away from you. Trying to push them up an alley requires that you get so close behind them that you are smack dab in the middle of hoof-cannon range. At night, they are more likely to get run over, or at the very least delay you, because they won't get out of the way.

And at full pet level, they will run over you because they were so happy to see you.

Audra Brown has seen the dark side of gentle cattle.

TO SHIFT OR NOT, THAT'S THE QUESTION

To shift or not to shift?

For anyone operating a manual transmission, that's usually the question.

It is a common conundrum independent of vehicle type or configuration, and a choice that is relinquished to some distant engineer if you have an automatic gearbox.

In a tractor, the decision is largely based on how hard the attached implement is pulling.

Sometimes, for lighter loads, it is based on how fast you can stand to ride the rough ground.

For more finicky jobs, like planting, its less about how fast you can go, than it is about what speed seems to do the best job.

For heavy loads like plowing, its simply a matter of how fast you can pull it--based on the combination of horsepower, torque, how deep you run, and how hard the ground is.

The tractor is only capable of so much, and a gear higher will stall it or break something other than the ground.

In a pickup, it depends on the engine and the terrain. A good, high-torque-at-low-rpm motor means you can pretty much shift when you want. You can lug out of a high gear and putter up a hill as needed. The old naturally-aspirated Ford (via International) diesels of the early nineties are perfectly suited to a manual transmission for this very reason. A big, long-stroking gas engine is pretty flexible too.

A newer, smaller, engine-that-gets-to-the-torque-at-a-much-too-high-rpm, is not quite so great when equipped with a stick. If you were just running up and down a flat highway, maybe, but that's not really what pickups are supposed to be good for. In this part of the world, sand-crawling can be a more desirable feature than good highway mileage. Personally, in my pickup, I stick with an automatic.

Now, in less duty-driven motor vehicles, I like me a good 'ol stick shift. Skeptical looks sometimes ensue from my less gear-headed, more urban, friends when I declare that I prefer a stick shift in my cars. It's not because they are cheaper, its not because of better mileage, it's because it's more fun. Driving around town is easy, mechanically speaking, and without the challenge of difficult terrain or a heavy load, the challenge of deciding when to shift is an entertaining distraction.

Audra Brown errs on the side of acceleration over gas-mileage.

THE TRAVELING FARMER SEES NOTHING BUT GREEN

There is a lot to see in the world, and traveling is one of the great pleasures. It is always revealing to see places and people and things that are not the same as what you see every day.

Travel farther and you can observe the great differences in the world, as well as the remarkable similarities.

It's not hard to find marvels of construction and wonders of art, especially in the more urban destinations. These things are magnificent and I recommend them fervently, but there are other wonders out there that not everyone gives the proper attention.

Between the cities and the monuments, usually seen on the journey between destinations, is where you will find the high points in a farmer's vacation.

(I know that I just used the term farmer and vacation in the same sentence, and I know how little that makes sense, so let's suspend our disbelief for a few moments, and if need be, consider this anecdote in the hypothetical.)

"Look at that [insert crop here]!" is perhaps the most common phrase heard when farmers travel.

If on a trip through northern Italy, the farmer will happily stare out the windows of the train, perfectly content to admire the agriculture that goes by.

There are familiar crops, in either a better or worse state than at home, and there are strange crops, admired for their novelty. All are viewed with a critical eye and a peculiar glee.

Eventually, though, the fields run together and even the farmer can grow tired of his fascination. At that point, equipment becomes of interest.

Tractors of unthinkable colors and bizarre configurations can be found as well as familiar machinery that seems unbelievable far from home.

Sometimes, and these are high-points, you get to see a combine wade through the tall, thick, beardless wheat of and English field, or see that poison-green grass of Switzerland get raked and baled and ready to feed all the cows with bells on.

Sometimes you see a tractor in traffic, surrounded by tiny European passenger cars.

You always see something different juxtaposed with something familiar and you'll hear about lots of green fields from the traveling farmer.

A VERDANT CONTRAST TO A BEAUTIFUL LAND

The natural geography and landscape of a place is a defining aspect of identity. In the wide west, it is easy to forget how different topography can be. A recent visit to verdant Alabama demonstrated a definite contrast to Eastern New Mexico.

The reliance on visual navigation common to the plains is not the most useful in other lands where trees and hills can hide even the biggest Bass Pro Shops.

After turning in at the sign off the highway, and a couple of miles of one-lane asphalt through the overlapping woods, I was beginning to wonder if the store would ever appear, or if this was just one very well-designed redneck trap.

It turned out to be nothing more than the longest parking lot driveway I have yet encountered.

The very range where we spent most of our time was surrounded on all sides by residential developments and the outskirts of the city of Birmingham. It seems quite likely that were the concealing trees to suddenly turn transparent, there would have been houses well within what a marksman of the plains would consider reasonable shooting distance. And did I mention the train running right behind the berms?

The traffic was denser than small-town me is used to, and it was fast, but for different reasons, I think, than here. In these parts, in my experience, the speedier end of driving is encouraged by the long flat highways and frequent isolation of any one motor vehicle outside the city limits. Indeed, the thought occurs that if I'm the only one out here, how likely is it that the one other car I encounter will have a set of flashing red lights?

When it is, of course, there is no one else to blame and plenty of room to get pulled over.

Down there, however, it looks like to me that there's not much of a place to pull anyone over, and too many ups and downs to get a clean shot with a radar.

The ground not being flat, and there being a lot of waters as well, the roads are all built up to go over and around.

Barditches do not appear to be applicable alternative parking spots.

I enjoyed the green and the people down Alabama way, but I'm still a flatlander that's glad to be home.

Audra Brown likes beef in her barbeque.

WHEN IT RAINS
VS.
WHEN IT BLOWS

When it rains, the proper and reasonable course of action is to immediately call your friends, family and neighbors to ask them a critical question.

"Hey, it's raining where I'm at. Is it raining there?"

The person whom you are querying might only be a short distance away, yet the question is still meant in all seriousness.

It is perhaps an indication of just how spotty high-plains rain showers can be, that the next neighbor over might, in fact, not be experiencing rain.

No matter how hard it's coming down outside, we just can't help but be suspicious and doubt that it could be as widespread as we hope.

Even after a beautiful winter of wet-weather, snow, and a suspicious lack of sand in the air, I swear that I called at least one person to see if the moisture-from-the-sky was also occurring in his or her vicinity on every occasion of wonderful weather.

I'm not overly fond of the phrase, "too good to be true," but that doesn't mean it isn't.

On another note, March has arrived.

Better never than late, it showed up in full force this week with a serious dose of high plains reality.

As an observation on the other side of my rain-doubting psyche, I walked outside and saw that the blue sky had decidedly been obscured by the brown of blowing sand.

I scrunched my face a little, unlocked my pickup, got in, and drove off towards home (better known as the darker parts of the brown cloud.) And what did go through my head but, "Huh, it s finally March."

I did not for one moment doubt what I could see.

I didn't call my family, friends, and the other numbers in my phonebook to see if it was blowing where they were.

Can you really hear that conversation without snickering?

"Hey, it's a sandstorm here. Is the wind blowing where you're at?"

Audra Brown is just glad March didn't come early.

WHEN THE SAND RATTLES

Some of the most interesting stories you'll hear told over a stiff cup of coffee are concerning the many and varied encounters a person has had with a rattlesnake.

The first thing you need to understand about rattlers, is that there are plenty of 'em and being aware of that, folks generally come prepared.

You keep a high-end, anti-snake device such as a shovel, a hoe, a shotgun, a pistol, a knotted rope, or so forth handy when you're out and about.

The next thing you need to know is that you rarely encounter a snake when so well-prepared.

No, you find a snake on that one day every so often when you didn't have enough coffee that morning and somehow left the house with pockets empty and the shovel "borrowed" out of your pickup.

You let your buddy try out your gun the day before and forgot to restock your pickup with ammo. And on a really bad day, your significantly-neater other decided to "clean out" the back of your pickup.

One way or another, you're facing a rattler with something that is in no way designed or had ever been previously considered for the purpose of snake-control.

If you are in your pickup, a good first bet is to pull up beside the snake, climb in the back and see what kind of heavy objects you can find.

Large rocks, hi-boy jacks, mineral blocks, even hammers and also-hammers (large crescent-wrenches, etc...) can be dropped to neutralize the target.

If you are in the town-rig or your pickup is clean, or you missed enough that heavy objects are not an option, you start looking for length.

The fact is, not-getting-bit directly correlates to not-getting-close.

With this in mind, you might pull out the booster cables, a log-chain, fence-post, or as it might happen, a wooden practice sword.

Use the floppy tools in a repetitive whacking manner, and the stiff tools to pin the head.

(Note that you should refrain from pulling the tool towards you when executing the repetitive-whacking maneuver. Vertical motions only. With the pinning technique, additional disarmament may be required.)

So, keep your ears open and your shovel close.

Audra Brown did not borrow your shovel.

BAD TRANSLATION, GOOD ADVICE

Beware the badness of the terrace.

Good advice from a book on the art of fencing with the short sword that I would pose was rather badly translated from French. A better translation might be, "Watch where you step." But the odd one seems to imply more of a sense of danger, and on the farm or ranch, that is the sentiment that needs to be implied.

What you accidentally step on can kill you.

Just one day this week, I know one house of farmers who found eight rattlesnakes.

That's a lot of snakes to not step on.

Every day, one must be on the lookout for those slithering varmints underfoot, and under anything else.

Another common badness to be found on the ground (and preferably avoided) is not so deadly, but notably unpleasant nonetheless. It comes in a range of green and brown hues, and a variety of viscosities.

Manure, barnyard, or call it a cow pie, no matter how you couch it, it is smelly, nasty, and it is neither a simple nor pleasant task to get it off your shoe.

There are places where it cannot be avoided, but even then, when you must step in it, it's best to know the specifics of the pie in order not to unexpectedly slip, fall, or slide. At least ice is seasonal.

Manure is all year-round.

Not just your own feet either. Every piece of equipment has a footprint; places it can go, and things it should not run over.

On a horse, there are four feet to monitor. One rodent-den of a hole and flying through the air to land in a pile of grassburrs is the least of your worries. The horse is probably worse off.

Now that I mention it, grassburrs deserve mention. Also cuckleburrs, devil's claws, bullnettles, and goatheads.

There are a lot of things that can trip you, tangle you, bite you, stab you, or just plain make the rest of the day unpleasant.

So, mind the gap, watch where you step, look for snakes in the grass, look for snakes not in the grass too, and beware the badness of the terrace.

Audra Brown warns the terrace to beware of her, but wears thick-soled boots anyway.

FARMING WORD PROBLEM

The average American household owns approximately two cars and a third.

The average farmer's driveway might look reasonably average. A pickup per person, likely a large SUV, and maybe even a town car or, if they are so inclined, an old cool car for fun.

Oh, wait, that's already a bit over the number and we're just getting going here.

For calculations sake, let's figure this for a farming family of four.

The bare minimum would be four vehicles designed for personal transport, but even six would be admirably low. Then there are tractors, say about eight. Some standard row-crop Deeres, a big 4WD or two, a loader with hay forks, and the one that we still call new.

Then there's a grain combine and maybe even a self-propelled swather (crazy wheels and all). ATVs, motorcycles, and a lawn mower. I lost count already, but let's call it twenty to make it nice and round (and we're still giving a low estimate here.)

Have you ever seen that puzzle game with three posts and a set of graduated discs?

The discs start on one post, stacked in order from big to small. The goal is to move them all to another post, without moving more than one at a time or putting a bigger one on top of a smaller one. Two rules.

Ends up taking a lot of moves to get the job done.

Moving equipment and vehicles on a farm is very similar, but worse. Let's consider a simple one tractor problem.

If the tractor you need is in one place, and the implement it needs to be hooked on to is in another, and home is elsewhere, how many vehicles does it take to get it done before supper?

You have to get to the tractor, and then to the plow, but you probably need a pickup for the tools and air-compressor at both locations. Then, hooked on, you need to be in yet another place to get the work done, and eventually, you have to get back home again, but you won't be taking the tractor. How many drivers do you need? And what order do you move them all?

Now, in my experience, there's usually more than one tractor, you probably needed to unhook and put away some implement first, and all the places are miles from each other. Obviously, it can get real complicated quick.

Audra Brown says there will be a test on this later.

WORK CLOTHES

There's an interesting expectation among those who don't work in the dirt for a living, that everyday clothes are relatively nice and clean.

I remember when I finally noticed the glaring difference in unstated expectations regarding what "work clothes" means.

To me, work clothes are the bottom of the heap in terms of repair and shine. Britches with holes honestly worn out from the rub of my pocket knife, with the knee burnt out from getting too close to the searing iron that time, or with a unique and artful collection of holes that only battery acid can create.

Shirts with logos I don't like, or plain and bought off the bargain rack; anything that I don't feel bad about losing. Because, any piece of fabric worn to work could be irrevocable destroyed at any time.

I finally threw away an otherwise perfectly fine shirt for no other reason than the gear-oil smell was so strong it gave me a headache every time put it on, even after a year of frequent washing.

Some stains and scars just never go away. Gear oil, epoxy, primer red paint, silicone, burns, breaks, and barb-wire tears...

Other eye-sores and scents can be washed away, but will provide a marked increase in your social space should you venture into town on the way to the shower.

I usually try not to go to town in the more severe cases, despite what amusing reactions I can imagine people having upon observing me after a long day branding cattle that had more than a few small horns.

Dirty, blood-spattered from my hat to my boots, dead-eyed and suffused with the smell of burnt hair and bovine excrement, even I get a little upset looking in the mirror after praciticing the Way of the Head.

I have no idea how many times people must have said something to me about work clothes and meant something very different from what those words mean to me.

Now, I translate that to mean clean, casual clothes, not for work and not so spiffy as what I ought to wear to church on Easter Sunday.

Audra Brown had pocket-knife holes in her britches before you could buy 'em that way.

FRIENDS IN LONELY PLACES

They say it's "who you know" and that's the truth. It's also where you know them.

Friends are one of the most irreplaceable resources in life.

Friends who live in the town three miles from where you just broke down are really hard to replace.

This story starts with a car that lived in the dead center of Fort Worth. Find a map, look under between the words "Fort" and "Worth." You'll see two lane-happy interstates that intersect in a mess that looks like it'd take more than WD-40 to untangle. Right there is where this car lived.

These days, it lives much closer, but before it could relocate, it had to drive the few hundred miles in between and the driver had to get a ride to Fort Worth. Luckily, the driver convinced a kid with a land-yacht to make the trip. Getting there, meeting the car (and it's former drivers), and even getting out of the metropolitan area on a holiday weekend in a new car and a land-yacht was easy. Only a few horns were honked and no one made the mistake of scratching the land-yacht. A nice Texas highway with little traffic was found and home was only a few, straight-line hours away...

But when it seems too easy, it usually is. The adventure became notably more adventurous about halfway back.

Three miles from a town with one parts-store, well after hours, an idler got thrown and the fan-belt followed.

One phone call led to another and providence smiled. Someone known lived nearby. That number was found, and guess what? Out of town. And the friends' friends who lived even closer? Also out of town.

But hope is not lost when everybody in town knows the owner of the parts store.

Bubba could not be reached that evening, so plans were made to stay the night in the least-not-recommended motel in town. But friends can be so nice that you can't ever really thank them enough.

They can let you stay in their house, that you've never even been in, while they are gone. Then, the next morning, they can give you Bubba's cell number. Things work out in the best ways and friends make it happen. Sometimes it's too good to be true.

Sometimes, Bubba is closed because he's gone dove hunting.

Audra Brown had to drive a little further to get parts and hopes the dove were plentiful.

AN ODE TO THE AVON LADY

Country girls can get work done and look good doing it.

Now, I'm about the farthest thing from an expert on the subject. (My version of that motto would end with: "and read a book while doing it.") But, there are more than a few ladies who do it all and do it right.

One of those women, whom I admire, was the Avon Lady.

Before there was UPS, before there was supermarkets and discount department stores, way on back before there was internet, the folks who didn't live particularly near the city center, had to work a little harder to procure those products that they couldn t grow or make themselves.

Lucky for them, the Avon Lady passed their way sometimes on her way to check cattle or maybe haul a truckload of grain.

She didn't usually have the time to stop for long, but stop by she did, and she always had a book of colorful delights...and sometimes she brought samples.

The best times, though, she brought stories.

Orders were accepted, and delivered the next time she had some work to do that caused her to pass near enough by your house.

The Avon Lady was a fixture, a common, memorable event that I know touched at least a couple of generations and served a spread-out community for many, many years while getting a whole lot of ranching and farming done in the process.

The Avon Lady did eventually retire, and to be honest, I haven't thought a second about the catalogs, or the goodies that she used to bring, because that was never the prize.

The Avon Lady was the delight.

She had stories to tell and adventures to relate. She'd have a cup of black coffee and pour you one too.

She was an amazing person and an honor and pleasure to know.

Whether the need was Avon, entertainment, a cup of coffee, or some good help chasing and catching a calf, The Avon Lady had you covered.

Audra Brown will miss the Avon Lady.

THE COUNTRY HOSTESS

There are a lot of hats for people to wear.

In the country, there are fewer heads to fill those hats.

Some hats are harder to find and that just makes the few and far between who wear them all the more notable.

One of those rare hats is hard to describe. Party-planner, host/hostess, are both accurate but incomplete. In a futile attempt to describe the essence of the position, I'd call it the one-who-arranges-get-togethers.

The term get-together is more accurate and meaningful than party. It implies the meeting of folks who know each other, and in truth, the meeting of folks who ought to know each other. That is the true gift of the country hostess, that knack for knowing who to put together; the ability to make a get-together something special that left you permanently better off for being there and visiting with the other folks that were there too.

It's a hard hat to fill. Do you know how hard it is to convince far-flung farmers and ranchers and other folks who live only a bit farther from town than they do from each other to come to a party?

They live spread out for a reason.

Social gatherings are not opposed, but they are frequently overlooked or overridden by seemingly superior priorities.

A person who can consistently gather these sorts of characters, not only once, but to the point that an invite to her parties was something more than just another social occasion?

A person like that is something special.

I'm certain that this archetypal mastermind of meetups has been a treasured fixture for many ages. When the invitations were delivered horseback, and the distances were even harder to cross, that person put on the get-togethers that introduced friends, future mates, and important networking connections.

There is something magic about such shindigs that the marvels of modern social networking, for all their amazing breadth and possibility, cannot quite replicate.

The great country hostess is determined, wise, and tireless, for anything less would not be capable of making the wonderful get-togethers happen by bringing together the busy, distant, and just plain stubborn.

Audra Brown tips her hat as one of the great ones passes.

ALL IT TAKES IS COFFEE AND A BARN

I talk a lot about the trouble that comes along on the farm and out on the ranch.

There's a reason for it.

Trouble is interesting, and conflict is the core of any story, including the ones I get to tell here every week.

However, there are moments that are completely devoid of trouble and conflict, but are no less for that apparent lack of discord.

The 31st of December, I lucked out and found one of those times.

It started with a barn, and a little sister with plenty of gumption. We moved the many sack-stacked pallets to the edges of the room, swept between, and put out a couple dozen chairs that we scrounged up from somewhere. There was a couple of rugs rolled up (ostensibly, in-storage) and an old coffee table. We put them out, dusted them off and before you know it, it was just plain cozy.

As you may recall, that night was rather chill, but we had that covered with a big portable heater we'd gotten to warm things up when we are working outside in the colder parts of winter.

To be honest, the cold was nice, because that heater works really good and we might have all roasted otherwise.

I fried some chips and made some queso. There were leftover pies from the week before. We also found some unclassified peanut candy (don't look at me like that, it tasted good), some rice crispy treats, and despite the fact that it wasn't at all required, the friends and neighbors who came brought goodies as well.

Add an extra-large pot of coffee, iced tea, and spiced cider.

We all gathered up and pretended that the heater was a fire (it was sure enough hot as one, whew!)

I've awaited the new year in quite a few different ways, some keep you up, and others don't keep you awake well enough to last 'til midnight.

I'm gonna go so far as to say that nothing is more effective and fun than bringing in the new calendar with family, friends, and neighbors in a warm, coffee-stocked barn.

Audra Brown didn't resolve much.

REASONS TO CELEBRATE

When you work all hours of the night and day for weeks on end, you're always looking for a good reason to stop for just a little while.

Sleep requires a cessation of labor, but other than that, justification for long pauses are hard to come by. I mean, it's hard to find a really good reason to stop and take a break.

That's why it's best to be very liberal and open-minded about what constitutes a good reason.

Company doesn't always justify a long stop, but at least long enough to slow down and let them ride in the tractor with you.

People always want to know "when's the best time to come out and visit?"

They want to come when we aren't busy.

That time does not exist. We are always busy.

Therefore, the best time to visit is when you show up. That gives us a reason to be less busy for a little while.

The point is, we're always looking for a reason to slow down and celebrate something.

Got a new tractor?
Put a little dent in the note?
Shipped a load of cattle off?

All these things are at least a reason to stop and have a cup of coffee or a glass of iced tea.

While most of these impromptu (and quite a few more planned events) occur in the field or in the barn, they are no less enjoyable, and in my view, sometimes more fun than a get-together in a more civilized location.

The last party that happened was in a barn.

I learned about it just in time to bring the pizza, and arrived just in time for the main event.

It was an engine-starting party.

A new block, all put together and turned over for the very first time is not just an acceptable reason to have a party, but a great reason.

It sounded great, it was suspenseful and exciting, and even better, it was a success!

My advice is to always be on the lookout for something to celebrate. Life's an adventure, not a trial.

Audra Brown wishes she had a new engine to start.

THREE BIG NIGHTS, ONE LITTLE TOWN, INCALCULABLE ENTERTAINMENT

Three big nights, once a year, every year.

For over sixty-five years, the rural community surrounding Floyd, New Mexico has ponied up an impressive roster of talented musicians, singers, poets, and a few unclassifiable acts of entertainment. Not all the members of the Country Jamboree are from Floyd, in the strictest since, but most are from the extended community, at the very least. And rural communities can extend quite a fair distance...

As spread out as farmers, ranchers, and other residents of the rural plains tend to be, bringing any significant portion of them together is quite a feat and rare is the event that has done that very thing, for such an impressive length of time. Music brings people together, so they say, and it appears to be a legitimate theory.

Not only does the Floyd Jamboree bring in the rural residents, it brings folks that live a little more close to one another and gives them a chance to come out to the markedly remote and tiny municipality of Floyd, and for three big nights, become a part of the rural community.

It's a good time for everyone who enjoys a little music.

There's no drinking but for coffee and cokes.
There's no cussing, except maybe in a song
(and cussing in a song is different, as we all know.)
There's no fighting, unless I get to participate
(and I usually don't have the time to spare.)
There's no spitting...
Well, actually, that is something I cannot guarantee...

Seating is first come, first serve--same seven dollars for
any seat in the house

Hosted by the Floyd Lions Club, usually on the last full
weekend in March, the proceeds from the show go to
charitable, Lion supported projects. The band gets paid
in coffee and cinnamon rolls, and the other entertainers
get what the band doesn't find.

It's a fun time, and surely worth a drive out to Floyd. At
seven bucks, you get your money's worth in the first five
minutes--and the rest is free!

*Audra Brown will be behind the sound board, and on
stage, and behind the stage, and in between...*

BARNS ARE THE BEST CAVES

"Man Caves" are in.

Everywhere, there is decor and references to this relatively recently coined label for what, in olden times, was probably known as the garage.

A place to go where you can enjoy the company of your friends and look at cool stuff; a place somewhere between a house and the outside; sheltered, but undomesticated. I'm a fan of such a thing, but I think that one more important idea is that it is place to do cool stuff--and by "do," I mean "work-on."

And this is where I think that the modern man-cave seems to be lacking based on the popular versions I've seen. A garage is closer, but what they really want--and don't realize it--is a barn.

Let us contrast. The modern man-cave is festooned with images of hot rods and muscle cars, designed with colors, trim, and accents that invoke rusted and rough-worked metal, bolt-heads that don't hold anything together, weathered wood, and unusable simulacrums of tools.

Now, consider a barn. Instead of photos and models of sweet automobiles, one can gaze at the 1:1 scale reality.

One can smell the lingering scent of burnt exhaust, oil, grease, rubber, and anti-freeze. One can sit and listen for the drips of leaking fluids, or even better, the silence of good seals. Walk on a concrete floor patinaed from the many puddles of oil, sweat, and anti-freeze.

Polished by the rolling castors of a creeper and the grit of many passes of oil-soaked floor-sweep. Instead of metal decor, there are scraps of it, lengths of it, sheets of it, that await the next project. There is not tool-themed decor, just cabinets, hangers, and racks of the real thing. Wrenches, drivers, saws, and if you're lucky, some special tools like the one it takes to put a steering-wheel back on. One fridge is full of bearings, and wheel-bearing grease. The other fridge actually works and its for when you get thirsty working on stuff.

Too much work, not enough entertainment you say? Not so. With the skills and tools so handily near, you've hoisted an antenna and hot-wired an auction-bought projector screen. The Super Bowl never looked so good as it does in ten-foot HD. All your friends, tons of food, a big screen, and room to play pool, air-hockey, foosball, and poker and still have a view.

Folks who don't have one, want one, and those that do have one, need another.

Audra Brown will be watching in the barn.

125

CATTLE WAIT FOR NO TURKEY

Thanksgiving is one of my favorite holidays. Good food, good company, and (hopefully) a good football game or two.

Aunts, uncles, and all my cousins get together at Grandma's for a carefree day of eating, playing cards, and catching up on what we might have missed.

You know how it goes, after we stuff ourselves at least twice, we all sit around and trade stories until we feel like there might be just enough room to fit in another plate of pie.

Or, that's how its supposed to go...

Seems to me like Thanksgiving has had more than its fair share of catastrophes and adventures.

And just like any supposed day-off in the world of farming and ranching, it's only off until something happens...

...and usually something does.

One of the biggest incidents I remember was just a few short Thanksgivings ago. I'm not sure that we even made it to Grandma's before the call came in. Grab a turkey leg, something suitable for chasing cattle, and get going.

My uncle had just turned out a bunch of calves on wheat.

They were what we call Florida-heifers, which, in short, means that they are scared of neither man nor gator and crazier than...(I was looking for a good analogy here, but dern if I can't come up with anything crazier this side of Hollywood.)

Anyhow, something had startled these already-addled bovines more than a hot-wire fence and they scattered all over the place, and off the place, and nearly to town in some cases.

I've never seen a herd of cattle run off in quite so many directions or get so far from where they are supposed to be in such a short amount of time. We spent the whole day trying to locate and retrieve those dad-gummed heifers and in case you're wondering, a bluetooth headset is mighty handy on a four-wheeler in such a situation.

Eventually, after an exciting adventure and more hijinks than can be properly conveyed, the cattle were returned, the fence was rebuilt, and the turkey was consumed.

It was quite a normal holiday.

Audra Brown is thankful for adventure.

THE CHRISTMAS AMBUSH

Christmas is a time for stories and friends; food,
faith, and family; generosity and thanks. As my story
lengthens, I find that the giving of gifts is far more
my favorite than the getting of things. So, here's my
Christmas gift to you, a fun little fiction about some
holiday havoc that while untrue, is certainly set on some
fictional farm. For your amusement,   The Christmas
Ambush.

Twas the night before Christmas, and just so no one
gets shot,
There's no front door delivery. We keep the gates locked.
Down the road we must go, a few miles and more,
To pick up the presents delivered to Grandma's back
porch
After some coffee, good and hot and black,
We said our goodnights and then we made tracks.
It was just about then, on the way home,
I heard a western tune playing, so I answered the phone.

It was my dear mother, on her way home from town,
Said she needed some assistance to chase someone
down.
Of course we were packing, my brother and I,
Not to mention little sister, (we just call her Deadeye.)
Grinning at each other, eyes shining bright,
We drove on more slowly and turned off the lights.
Then there it was, parked at our gate,
Some kind of red vehicle, unknown model and make.
We waited for word that we weren't alone,
We had 'em surrounded, no place to go.
Rolled down the windows, freed up our guns,
Turned on the lights, and moved in as one.

There in the headlights, you could see pretty clear,
There's a big ol' red sleigh and some strange lookin' deer.
I guess that the driver was takin' a break,
But when we showed up, he came right awake.
Before we could get out, and ask that old man,
Just what he was doin' on our patch of sand,
He started a hollerin' and callin' some names.
Them deer sure paid attention when he popped those big reins.
I reckon it's a good thing that our fence ain't too much higher,
'Cause short as it was, they still broke the top wire.

Early next morning, underneath the tree,
There was a roll of brand new barbed wire, addressed to me.
"I meant no offense!" read the short note.
"Here's for the fence that my sleigh broke."
There were quite a few presents, more than we'd ever seen.
Pliers, parts, and ammo. And some new guns to clean.
Now there's a bit of Santa's list that's permanently marked.
When you look up my family, I'll bet naughty is crossed off.

If there is a lesson in this story somewhere,
It's that next time we'll call our uncle who can cover the air.

Audra Brown wishes ya'll a Merry Christmas.

129

GET YOUR GOAT

When life gives you goat milk...bake a cake with it.

And make delicious cheese, and a homemade, vacuum-operated, goat-milking device.

That's what a certain kid (of the non-goat variety) did and I'll tell you what, it was pretty tasty. The kid now has one, blue-eyed, well-horned, king-of-the-haybale, fluffy, wool-endowed buck goat and one blue-eyed doe with a very impressive beard. The female is the newest member of the kid's menagerie and seems to be fitting in quite well. I believe that the new goat is even aware that it is, in fact, a goat. Perhaps, with the goat population increased to two, the other goat will get a better grasp on his species identity. I believe he knows he is a goat, but is unclear as to what that means.

Some animals have much more serious identity issues.

There once was a baby Holstein heifer who was saved and nursed to health from a premature birth and is now four-feet tall, has increased in mass approximately twenty-fold, and is now an 800-pound puppy. Having far outgrown the backyard she grew up in, it was time to send her off to meet others of her own kind.

They were not dogs, much to her surprise.

Her only previous encounters with such cows were chasing them out of the yard with the other dogs.

Her best friend was a cat who thought it was a cow, and her roommate was a goat who just thought he was better than everyone else.

Being trapped in a corral with other heifers of like age was apparently uncomfortable to the point of great amusement for those who got to observe.

The heifer took a page out of the Vulcan-Chihuahua's book and refused to look at the other heifers or in any way acknowledge their existence. She followed her mother (a human) around until that was not an option.

One of her new roomies was not so shy, though, and decided to say hello. This new friend moved towards the heifer's rear (she was still not looking) and as she approached, the heifer-who-doesn't-know-it started walking away. It was as comical as it sounds as she continued to increase her pace gradually (and without looking) until she was actually running away to avoid socialization.

It works out and friends get made, but it's hard to be a heifer if your sisters are all dogs.

Audra Brown likes that goat cheese.

A FIELD HAIRCUT

After you roll your hair up in the wheels of the floor creeper while working under a car or tractor, short hair begins to look very attractive in the sense that it would cause less inconvenience and pain.

I certainly rolled mine up in a creeper one too many times and after the last one, it wasn't a week before the hair I'd kept long for most of my life got chopped right off and proceeded to get even shorter over the next couple of years.

Short hair is light, well-ventilated, easy to manage- -and unlikely to get wound up around any rotating mechanical devices.

The downsides are that I can't use it as a weapon in sparring and it needs cut quite often.

This means that I sometimes can get a bit desperate when it starts to poke my eyes and strangle my ears.

In the midst of what is a busy schedule, finding the time or even remembering to try to find the time to get my hair cut is a trick that I don't pull off very well.

Instead, I often end up taking my hair-hat to the one shop that will cut it at odd hours and a discounted rate.

The farm and ranch, full-service salon run by my kid sister.

Get your head trimmed, your wool shorn, and your bovine's ears marked in one convenient location.

Sitting in a chair in the yard, she starts by wetting down the hair with a garden hose. Then, blades fly and so do a few flies as she starts the trimming.

In the meantime, the subject of the experiment--I mean, the client, which is me, collects the attention of all the ranch dogs who gather around to watch.

The Evil Cat and her kittle come around to see what is happening (though with much more skepticism than the hounds.)

The yard cattle chew in the background and before long, the sound of goats heralds the arrival of yet more animal observers.

There's no mirror, no refund policy, and not much say in what your final hair-hat is going to be, but it's an adventure every time and so far, I'd have to recommend the results.

Audra Brown figures worst case is it all comes off.

THE SEED OF FASHION

Fashion follows the bold.

A seed of style is planted and if circumstances favor the trend, it grows into a feature of folks' attire that is copied and worn with little regard or knowledge of where it all started.

It is true that some fashion evolves or is brought about by odd and non-functional causation, but it seems to work the other way as well. There are many times when the wear and tear of life in the field of agricultural production left it's mark on my attire and I was unwittingly fashionable.

Pants are protection from the many objects, animals, and substances that lurk in day-to-day business of ag. They are not acquired with fashion in mind, but with consideration for how resilient they might be to getting kicked by a cow, getting hung up in the fence, or the inevitable encounter with a caustic chemical or a burning brand. The best fit has nothing to do with showing or hiding your form from a cow, but how fast you can run, how well you can climb, and how high you can step to catch that stirrup. To buy my clothes with pre-made holes and no room to move would be unreasonable, but I find that fashion strikes even the best of britches.

A hole in the knee from a red-hot searing iron that I kneeled too close to on branding day, thankful for the fit that meant the fabric was not tight to my skin when it was burning.

An artistic spatter of small to pin-sized holes are the result of battery acid that was conveniently stopped by thick, un-perforated denim in lieu of my skin.

The hole up high on the front of the leg didn't happen so sudden, but is inevitable just the same. The one on the left is where the bulge of my pocket-knife eventually wears through, until the tip of my pocket hangs out instead of staying in. The right one is usually from my ring of keys. Neither are intended, but happen nonetheless.

Beyond the various pant-leg perforations that occur, there are various stains and permanent adornments that can also become accidental fashion. Paint, lube, epoxy, and light acid spatter or burns. There are far too many paths to incidental fashion to cover them all today, but I'd wager I've been mistaken for hip on occasion when really I just hadn't gotten around to throwing those pants away.

Audra Brown only wears original, one-of-a-kind tears, decor, and perforations.

STRANGE EQUIPMENT

Look, on the farm! It's a loader, it's a skidsteer, it's an ATV! No! It's. It's! It's...

I have no idea what it is, but it's either genius or the silliest thing you've ever seen.

This story begins, as many of the great ones do, at a farm auction.

Every sale has a certain amount of the same things. There's usually some tractors in various states of repair and color. Some trucks, trailers, and farming implements that may or may not still be of any use. Piles of wire, posts, and pipe of all sorts. Two or three mysterious stacks of tires that if they have a chance of holding air are worth spending a few bucks on. Might even put on one of those pickups that you buy hoping it'll run for a few months before you bring it right back to sell it on to some other hopeful soul.

The bread and butter is always there, but the fun really gets going when the more exotic items show up on the auction block.

Anything that is not standard farm equipment, yet can be described by the acronym, ATV, is always in great demand.

Four-wheelers that look like they've spent more time upside down in the manure pile will never go cheap, even if the keys are lost and the gas smells older than should ever be allowed to sit in a tank.

A more recent addition to the hot-item slate is the four-wheelers big brother that comes equipped with actual, feet-in-the-floor seats, a steering wheel, and a built-in rear container that can--if you're feeling generous--be called a dumping bed.

On the opposite end from the sporting vehicle is the ever-useful skidsteer. It's small and maneuverable with more available attachments than a lego set and can go where no loaders have probably gone before.

Every now and then, something equally cool, but amazingly strange also appears at the auction.

Like a rig that seems to be an adorable hybrid between a skidsteer and an large ATV. No one knew what to call it, some called it unkind names, and one farmer looked at the bizarre vehicle and thought, "Now, that's just the thing I need."

Audra Brown likes to buy stuff that is so strange...it just might go cheap.

MANY FRIDGES

The preservation of food is an important part of living out where town is a place that you don't get to often, and the grocery store is a place that you visit with a large, empty vehicle.

Take just enough help to push that second shopping cart, but the rest can stay home and keep the back seat free for the sacks of stuff you want to keep less dusty.

You might live in the country if the dry pantry has its own navigational protocol and locations are given in terms of compass directions.

The beans go on the second shelf, north end, east side. Always have. The potato chips on the middle shelf, south end by the door, west side. The cereal is a couple shelves up from that and the iced-tea bags are one to the right. And yes, the beans take up two shelves.

In another room, or sometimes spread out across several are the freezers.

I'm not gonna even pretend there aren't at least two. You gotta have one just for the beef you pack in every year.

Even at dressed weight, that twelve-hundred pound steer you fattened out doesn't really leave much room for company in the freezer.

Then, there's the three or four bushels of roasted green chile that you still have to ration if you don't want to run out before next season's come in.

That puts us at a freezer and a half, at least, so far. That other half would be for fish sticks and frozen pizzas and the ham you got planned for thanksgiving dinner.

Or it would be, if you didn't need a place to keep the heads.

Have you ever reached down in the freezer, pulled out a plastic-bagged bundle that you hoped was that Thanksgiving ham? Or even the mountain oysters you washed up after the last branding?

But that bundle was not what you were looking for.

If you were lucky, it was just an antelope head from the last hunt or a ziplock full of snakeskins somebody shucked and salted and will maybe tan later.

If you're unlucky, you grabbed the pointy piece of a sharp-clawed former creature and ended up with something worse than a paper cut.

Audra suggests keeping the ice cream over the hides.

FEED THE DOGS, AND EVERY-THING ELSE WHILE YOU'RE AT IT

Feed the dog and bring in the mail.

Yeah, there s a bit more to it than that. When considering an overnight departure from the place, a farmer must find someone to check up on things while away. This job can't be trusted to just anyone, and just anyone probably wouldn't want it anyhow. It starts relatively standard. Take a look at the house, check on the dogs. Make sure they have feed and water, and hopefully they haven't had any close encounters with a porcupine. If said encounter has occurred, please remove quills as needed.

While you're there, feed the chickens, and the duck, and don't let the cats stay in the chicken yard...but the goat is okay. The goat is supposed to be there and make sure he doesn't have his face stuck in his wool. He doesn't get along so well when he's stuck on himself like that.

Pet the milk-cow-in-training, she thinks she's a dog, and she ought to have hay and water too.

Help yourself to the coffee pot and anything you can find in the fridge, but don't forget to grab the pitchfork and the pickaxe from off the porch in the winter---you'll want 'em when you need to break ice in the cattle and horses' water.

If it's cold, you'll need to take a look at the heifers' water over in the west pens and the cows' tank on east. Hopefully the ice isn't so thick. We'll assume the snow isn't so deep that you need to put out any more hay.

That's the thing about cattle, they don't need too much attention.

The rest of the time, when it's dry and hot, you'll also need to check that the sprinklers are moving and putting out enough water. There's two running on the west place, fields 6 and 2, and one over on the east farm, the one we call new. You know the ones I mean, just make sure they haven't stopped and they've got water all the way to the end.

There's a list of the pumps that are supposed to be running, in case the electricity blinks or it rains. If it does shower more than two inches shut 'em all off and the sprinklers as well, please.

There's plenty to forget to tell you about, so if you see anything else, just take a look if you would.

Thanks for your help, says the farmer, be back probably tomorrow afternoon.

Audra Brown knows how to break the ice

SOME CALVES ARE SMART, SOME ARE BUCKET-CALVES

Orphaned calves come about due to various circumstances.

Sometimes a cow dies before the calf is ready to be weaned, sometimes a cow is too stupid to take care of her calf properly, and sometimes, the cow is too old or too injured to make for a proper mama cow.

In these cases, the calf is gathered up and loaded into the back of the pickup and taken home to be bottle fed. (That's if you have help, otherwise, you'll likely have to let it ride in the front with you.)

If you're on a horse, you get to try to carry it over the horse's neck like in all those paintings and sculptures.

That's not too hard, but lemme tell ya, getting a squirmin' baby bovine up there on that horse is the trick. The cowboy-height-to-horse-height-ratio is a determining factor as to whether it's even worth trying.

But, eventually, even if you do have to call for a few extra wheels, the calf becomes yours to feed and fend for until it gets big enough to take care of itself.

These are the babies that become "Bottle Calves."

The first thing that baby calves need is milk. Typically, this can be delivered in a half-gallon bottle made just for that very purpose. Some calves know just what to do and get right to it. Some calves need a little instruction before they get the gist, but they get there too.

Some calves don't.

These I have had, and these I call, "Bucket Calves."

Calves who can't be taught to suck from a bottle makes one wonder just how much chance they would have had in their natural setting. Such suppositions are irrelevant, I know, but they do cross the mind.

Bucket calves, as the title suggests, must receive their lactose sustenance in an open-topped container such that they can slurp, sip, and spill the milk into their system without performing any complex tricks such as those required to suck a cow's teat.

Audra Brown says it's harder to carry six buckets of milk than six bottles.

MINI MILK COW

A certain kid I know has a new project, having recently acquired a brand spanking new, miniature milk cow.

And I mean what I say when I say miniature.

Despite the fact that I'm not into the dairy business my own self, I do know some folks who are, and one of the huge differences I've noticed between my side of the cattle business and theirs, is the babies.

When a beef cow has a calf, I don't usually have to do much once that calf is on the ground. There are occasions where I need to help a thing or two here and there along in the early stages, but once the kinks are worked out, the mama cow does the rest of the raising.

On the other hand, most dairy babies get to be raised up on a bottle.

I've done my fair share of that very thing, buying bull Holstein babies and bottling them twice a day when I was a quite a bit shorter than I am now.

I'm much lazier now than I was then, but this kid is not.

This kid has got the gumption and the patience to take care of the new project.

'Cause the new project isn't just a baby calf that needs fed and raised like any other, the new project is an under-sized, way-too-young, and tiny baby calf that needs a lot more attention than most can give.

That's why it is where it's at, because it takes a special sort of person to raise up something so fragile and this kid has done it before.

It takes an even more special sort of person to put in the time and energy, knowing that even with the best care, it may not work out.

I enjoy this kid's project because I don't have to do any of the hard work, but that calf is the cutest, softest little thing and it's getting taken care of by the best possible hands.

Lucky me, I get to admire the cuteness all I want without all the work. (Not that I'd claim to have the necessary patience, even as my shorter self.)

Someday, maybe I'll even get some fresh milk.

Audra Brown will not be doing the milking.

ONE-EYED CATS AND DOGS

When you're dealing with large livestock, heavy equipment, and never enough man-power, excitement can get pretty exciting on the farm.

Sometimes all it takes is a new barn cat.

I realize that the term "barn cat" is a little redundant, seeing as how that's where cats are supposed to be, but I also notice that some people don't have barns, and in the confusion, have cats in the house. Now, I'm not saying that there's anything wrong with that, I just don't subscribe to it myself.

My idea of a good cat is one that stays mostly out of sight and out of mind; a wily, sharp-clawed mouser; the kind of cat that you don't want in your lap; the kind of cat that is only domesticated in the sense that it's smaller than a wildcat.

The sorta cat that moved in our barn a couple weeks ago.

I've met a lot of cats, some more interesting than others, but, in recent memory, I don't recall one any more interesting than this ol' lady.

Word was, she was a fine example of a barn cat, and with two kittens in tow, she'd be likely to stay with us instead of run off.

The only other feature was a warning that she wasn't particularly fond of dogs. Not an uncommon cat-inclination.

They weren't kidding.

Got her moved in, showed her where the food and water is, and all was well. Then, one of the dogs came within twenty feet of her.

Say hello to our new, grey-striped, bob-tailed, one-eyed hellcat.

I've seen a cat hiss, claw, and get after a dog before, but only once the offending canine got in the cat's space. This ol' lady likes her space, apparently.

Before that dog could get closer than twenty, thirty feet, this cat took off running and Got Hold of that dog. She grabbed on and the dog didn't know what hit him.

Long story short, from the Pyrenees pup to the Great Dane, this cat don't care how big it is, or how close it gets. She sees a dog and she goes after it.

Quite a cat, she is.

Audra Brown likes this cat.

LIFE OF CHAINS

There are those items that a person should never be without.

Duct tape, a roll of blue paper towels, an assortment of nylon ties, some rust-penetrating spray lubricant, a fifteen-inch crescent, mid-sized pipe wrench, some Go-Jo degreasing hand cleaner, and a good chain.

Nothing can replace a decent length of flexible steel.

What else are you going to use when you need to pull someone, something, or some cow out of a marshy mud hole? What else can you boom down equipment on a trailer with that locks it down like you welded 'em together? How about when it's time to deep-break a field with the plow that takes two tractors to pull?

Tractor One hooks onto the plow and Tractor Two runs ahead and pulls Tractor One with a length of chain. That is a crazy job in and of itself, and a broken chain is dadgum scary in that situation, but it does the job.

What about a tow-strap instead? Well, for a few tasks, a strap can work, but straps flex and stretch and they don't clean up nearly as good.

For example, when a big, bloated, bovine corpse needs to be moved (either vertical or horizontally), a chain doesn't soak up the smell for long. After the next rain or a deliberate faucet visit, the chain is back to being unscented steel.

How about situations where a strap might get cut? Or when it just can't handle the weight? Lifting things that are too heavy to lift yourself, too big to lift yourself, and too awkward to lift with just a bucket or some forks.

A chain even has that unique ability to lock against itself to prevent all the slipping, sliding and shifting that could cause more than a little disaster. It also makes it most effective at pulling posts. Taking up everything from cedar posts to posts made of pipe or big well-casing, or even a tree. A chain just works for that job.

Chains live a varied life. They spend a lot of time waiting around in the back of the pickup too.

They drag dead cattle, hang critters up to skin, connect tractors, unstick things stuck in whatever, boom down loaded trucks, fall off trucks, get picked up out of the barditches, and pull lots of posts.

They lift engines and rocks and sprinklers and upside-down cows. They whack rattlesnakes and pinch your fingers and get tangled up in unfortunately-sized perforated metal grills.

Audra Brown has boomed a lot of chains.

THE 15" ADJUSTABLE WRENCH

Christmas time is here! It is the wonderful season of friends, family, and joy. It is also the season of giving. And if gift-giving is in your plans, I'm here to make my recommendations.

What do you get the person who has everything? Something that you can never have too many of. There is not one perfect gift, but there are things that are approximately close enough. Such a gift must be flexible in potential application; useful in a wide range of situations. We want a gift that lasts a long time and does not fail, falter, or accidentally insult someone unintended. (Indiscriminate insolence is sloppy. I suggest careful targeting.)

While more than one item fits this bill, it prompts me to recount the value of one tool in particular that I have found indispensable across a wide field of times, places, and problems.

There is no substitute for a fifteen-inch crescent wrench.

More generically, it is known to some as an adjustable wrench. No matter the moniker, the fact of the matter remains. It is the shiniest part of the hand-tool trio equivalent to the trinity of consumable necessity that is duct tape, nylon ties, and some sort of rust-penetrating lubricant.

The fifteen inch crescent is of an obvious length and weighs something like three pounds. Much lighter than a set of wrenches that would have to range from smallest to an inch and three-quarters, including the standard sizes, the sizes that are divided by 64 and 32, and the metric. In one tool, all those flat-sided nuts and bolt heads of average or better condition can be effectively manipulated with one tool.

But wait, there's more.

It can be used as a sort of clamp, holding flat pieces of material. With a secure hold and the leverage of the fifteen inches, you can grasp and bend a lot of stuff. How about those situations when delicacy and the civilized concepts such as rotation are just too refined to get the job done? Never fear. While not the perfect instrument of pounding that is a hammer, a fifteen-inch crescent wrench is admirably adept at pretending to be a hammer. It's the proper size that most cheater-pipes will shove on the end and add leverage. It could even be a weapon if the situation arose.

You can rarely go wrong with a fifteen-inch crescent wrench. Add a hammer, a pipe-wrench, and the trinity of consumables and the world is yours to take apart and put back together.

Audra Brown needs more of them.

THE JACK OF DAGGERS

There are certain tools that are hard to replace with any other. A hammer is a hammer...but a rock or a crescent, or quite a few other things can be used in a pinch to serve as a hammer, albeit of inferior efficiency.

In the other hand, we have a jack.

There are different kinds, but they all produce essentially the same effect, and do not have any other purpose to the design. That is to say, a jack is a device that you put between two surfaces, and it tries to push them apart.

The most common misconception is that a jack is only for lifting things up, even if that is perhaps a most common use.

Jacks can be used sideways, upside down, and all angles in-between. They are handy, irreplacable, and also frequently dangerous.

The true trick, though, is having the right jack for the job, because all jacks are not created equal.

Consider those little hydraulic ones that come stashed behind the seat of a new pickup.

They are barely adequate when conditions are best.

i.e. when you get that flat on level concrete/asphalt on a lovely spring day.

I don't know about you, but I rarely ruin my tires on such nice surfaces.

No, I'm usually a few inches deep in sand, in a bar-ditch steep enough to ski down, or maybe I'm just smack in the middle of a grass-burr infestation.

Floor-jacks, manual or air-powered are great in the shop, but that's as far as they roll. Portable air-jacks are nice, but require a bit more than just getting thrown in the back of the pickup and left there. That's to say, required accessories (air-compressor) not included.

Jacks in general are dangerous because they are usually holding things up/apart that would cause more than a little damage if they were released.

The most handy and similarly nefarious jack has to be the Handyman jack. Although it goes by many names, it is so useful that all the smashed digits, broken bones, and worse that we've heard of or experienced, doesn't stop anyone from using it all the time.

Audra Brown still has all her digits.

ONE MAN'S SCRAP

Even though it's still snowing, people are already getting into the "Spring Cleaning" state-of-mind.

There's a lot of decluttering advice finding its way onto my computer screen and there's some folks already in the midst of yet another round of clean up.

Hoarding, or the unreasonable compulsion to keep things of absolutely no value, is not something I would recommend.

Compulsive cleaning, I fear, is not much better.

In the process of farming and ranching, many things are built, broken, fixed, and everywhere in between. The result is that there is a potential accumulation of less-than-new parts and pieces. Admittedly, there are bits that are extremely unlikely to ever serve a useful purpose again.

A busted float, a tangle of wire after it's been removed from some unlucky rotating mechanism, bent bull-panels, and broken sickle-teeth come to mind as detritus destined for the terminal scrap heap.

In contrast, there are also a lot of bits that may yet prove useful, frequently in duties far removed from the originally intended purpose.

Indeed, there may not be a clear and obvious use for that ten-foot piece of rod from a retired circle-sprinkler, but is it really ready to be committed to the doom of the scrap iron barrel?

What about that piece of flat iron that was the short end of last week's project? It'll be just exactly the perfect thing for that project that will come up next week, or at least maybe the one after that.

There are definitive differences in philosophy regarding the virtues of tidiness and a Spartan workshop versus the benefits of variety and something that looks like my barn.

Whatever the position one takes on the issue, there are a couple of things that might be considered before throwing that extra material away.

If it's sound enough that it'll have to be cut up into considerably smaller pieces before it can be hauled off, maybe it's not junk after all.

If it's not yours, it's not yours to scrap.

If you search for things to get rid of, a re-examination of priorities might be beneficial.

Audra Brown would rather see possibility than the floor.

RUST WAITS FOR NO PLAN

Rust waits for nothing, not even a good idea.

It is difficult to fix something, if you can't find it.

It's probably worth more freshly broke-down than broke-down and been sitting for a few years.

Whether inclined or disinclined to clutter, even collectors such as myself have to take stock and toss out some stuff in order to make room for new possibilities.

As I write this instead of cleaning--unless you count the words I'm tossing out of my head--certain precious piles of future project material waste away in want of the time to tackle them.

Perhaps not every piece of scrap iron that collects in a pile behind the barn contains enough potential to keep around.

It even occurs that the revenue returned from hauling a load of the lesser pieces off every now and then might allow the purchase of new or at least better parts to fix something.

...or at least buy the coffee needed to complete another project.

Since money is a most moldable tool, it does seem reasonable to think that in lieu of a thousand special pieces that might someday prove to be the required shape, a few green pieces of paper might guarantee the proper piece next time a part must be acquired.

In fact, that sounds like a mighty steady bit of logic.

But debate is more fun than an answer.

So, is it better to have the pieces to produce what you might need in the nearby stack?

Or to have resources in your pocket to skip the line and go directly to the part number (do not collect the packaging, do not sort through 200 others)?

I for one, don't think there is an answer.

Both ways have their flaws and their features.

In the end, what does it matter, as long as the job gets done and the equipment stays running and the seeds get planted and the harvest gets sold and the cattle stay in?

Audra Brown doesn't throw away books.

PASS THE PETROLEUM PLEASE

Petroleum products rule.

They are awesome.

Don't try to argue. You won't win.

The author is quite familiar with the use of these marvels, and among their many qualities, they are frequently quite flammable. Fact.

But let us take note of some less-exciting, but equally important uses.

There is, of course, fuel. Gasoline, and the almighty diesel keeps the engines running and the parts a coming. If there is one certainty in the chaos that is farming, it might be the diesel bill. All year round, the diesel tank better be well above empty.

When business is slow, you might need a hundred gallons a week. Or, when the tractors are in the field, a hundred gallons before lunch--per person.

Comparatively, in a car that gets 30mpg and drives a hundred miles every day, a hundred gallons of fuel would get that car close to a month. Needless to say, farmers have a particular perspective when it comes to fuel consumption and we don't measure it in mpgs.

And fuel is just the highest-volume player. There is also the liquid lubricant.

Oil comes in many variations and every one serves a purpose. Engine oil keeps things free, transmission oil keeps things moving in the right direction, and penetrating oil makes rusty bolts and other suboptimal leveraging situations possible, giving hope for disassembly where there otherwise would be none.

Don't laugh. I'm not kidding. Penetrating oil is da bomb...figuratively speaking, of course.

Last of the great petroleum derivatives is my personal nemesis and friend, grease.

The less liquid lubricant.
The stainer of many a shirt, and the getter of in everywhere. It's the good and the bad and the ugly, and deserves a hearty thanks for many saved bearings and for all the bearings that went out and were kept turning anyway with the generous and frequent application of grease.

Before you head out in the morning, make sure the back of the pickup is stocked with a variety of oil, grease, and fuel jugs.

Audra Brown is definitely a grease-something, but not a monkey.

THE FARMER WHO HAS EVERYTHING

Thanksgiving passes with the consumption of mass quantities of turkey, stuffing, and far more importantly--cranberries. Friends, family, and hopefully a good game of football. For many, it also triggers the official start of the countdown to Christmas and the mounting consideration of one small aspect of that holiday---shopping.

Not the most important feature of the holiday season, but it is, for better or worse, notable. Whether you chose to brave Black Friday for the deals (or for the fights. I hear it s better than pay-per-view), or forego the chaos and take a more leisurely approach, the biggest conundrum is always simple. What do you get for someone?

I cannot say there is a universal answer. Gift selection is a moving target with far too many variables to nail down for long, but in case you have a rancher or a farmer on your list, I will endeavor to give a few thoughts that might narrow it down.

A special tool that does a common job ridiculously well. For anyone with t-post fencing to do, maybe a tool that handles the clips far quicker and more consistently than a pair of pliers.

A better filter wrench is always handy. And if you find something that makes cleaning off a sweep-plow any easier, send me one too.

One can never have too many 15" Crescents, Channel-Locks, or Gerber multi-tools. A cordless, powered grease-gun is a modern invention that everyone needs. Duct tape, nylon ties, and a good rust-penetrating lubricant should never be in short supply. A sharp knife is always good. Audio books for those long hours on the tractor. Caps with lights in the bill are great. Did I mention you can never go wrong with a good, sharp knife?

The thing about a gift that one can often use is that it will be remembered and appreciated over and over again. Novelties get forgotten, and consumables evaporate, but a good tool will be treasured year after year.

If your gift-target is too prepared and all these things seem redundant, I have one last idea that is unlikely to be repetitive. All those hours on the tractor, or those cold days feeding on the ranch, what would be better than a hot shot of espresso to wake you up or warm you up as needed? My pick for this year is a 12-volt espresso maker for the farmer or rancher who already has everything.

Audra Brown doesn't have portable espresso.

I CAN'T BELIEVE IT'S
NOT DIESEL.
BOOM.

Life on and about the farm requires a number of skills.
Some, quick to master, others more slow, and then there
are those skills that must be learned before it's too late.
There is no try, just do it right and don't blow it---or let
it blow up.

Case in point: Gas-in-jug, and the critical skill of
flammables identification.

Containers of gasoline, diesel, solvent, oil, and other
relatively low-viscosity liquids are often found.
Sometimes, they are even found when they are needed.
Rarely are they found with sufficient identification.

There is a theory, of course. Gas cans are red, kerosene
jugs are blue, diesel goes in yellow, and I know what
mine looks like, so I better not find it in the back of your
pickup, or I'm gonna get you!

Despite this not-quite foolproof system, confusion is
more the rule than the exception. After all, there isn't a
color-coded can for solvent, nor one for mixed-gas, and
is the mixed-gas for the chainsaw?
Or is it proportioned for brother's dirtbike?
Or is it for my oddball cycle that likes it lean?

And believe it or not, one does not always have a full set of colored containers in the field.

The most common style of jug is the always anonymous 'oil-jug' that unless it's still sealed, is very unlikely to contain actual oil.

Thus, the ability to accurately identify the contents of a said jug is a critical skill.

The smell quickly narrows it down to 'flammable liquid fuel' and in doing so, indicates the need to further investigate, lest it be an explosive mistake. The smell might also suggest a specific answer, but I for one, don't trust my nose to always be so precise.

If its red, then it's probably diesel and you can go from there. Otherwise, if it's more or less clear, continue to investigate.

Is it greasy? Oily? Kinda lingers? Or evaporates real fast? If you think it's gas, is it good? Or too old and bad?

Ask all the questions before you proceed and don't forget that diesel is oily, but so is chainsaw-gas.

Audra Brown throws the match from an arm's-length away, just in case it's not diesel.

PUNCHING COWS (AND SOMETIMES KICKING THEM TOO)

Audra Brown is more than a little awkward when writing a introduction to herself, which may be considered odd due to the fact that she considers herself very handy at most things--including writing.

She writes a lot, mostly fiction in the areas of not-romance. She's had several stories published, and has even been paid for one particular horror story. She is in the middle of the slow process of publishing her first novel and getting this book ready to go.

In the meantime, she writes more columns, novels, stories, and screenplays.

Writing is a lot like farming, lots of hard work and long hours up front, and then a long wait and a lot of praying, and then maybe eventually (or maybe not) the farmer gets paid. Farming actually seems easier sometimes.

Maybe that's why she still does it.

Though she'd consider her business to be more cattle-oriented, Audra is, and always will be, a farm kid.

She was driving dump buggies about the same time her contemporaries were starting school. She did not join them, instead spending her school years starting her own cattle operation and working on her family's farming and ranching operation.

Her education was unorthodox, and is still largely undocumented, but in no way lacking. Mathematics, science, and reading went hand in hand with diesel mechanicing, part fabrication, and all the other exciting problems that plague a farm.

Somewhere along the way, she made room for a few hobbies. Reading, leatherwork, competitive shooting, and muscle car collection and restoration--to name but a few. And one more that is perhaps one of her crowning achievements so far.

Martial Arts, specifically, Taekwondo, is a passion she has followed almost as long as she has known how to drive a tractor. Presently, she is a 5th degree black belt, 5-time member of Team USA, and repeating World Champion Power Breaker.

She's been around the turnrow, and around the world, and hopes to be around a few more times at least.

The important thing about Audra today, is that she is the author of this collection of reality-inspired anecdotes and hopes that you all enjoy the exploits and opinions that she will be sharing in this and future editions. Howdy ya'll!

THE RESOLUTION OF A CLIFF-HANGER

January is a big month. The start of a new year, lots of erased dates, and an extra column from me squeezed in.

Wow, I just realized that I've been writing this column for six months.

Dadgum!

It just seems like yesterday that I awkwardly introduced myself to all ya'll. I hope that somewhere in that time I've said something that you enjoyed or at least made you laugh a little. I know that I've enjoyed writing and I'm looking forward to all the stories I find this year.

Back in July, I mentioned some upcoming events of not insignificant magnitude. Namely, I told you about the local black belts headed to Rome, Italy to compete in the Taekwondo World Championships on behalf of the good ol' US of A. It goes against my writerly principles to leave the reader hanging...

...for too long. So, I figured it was time to reveal the end of that story.

Five local athletes went to Rome and five returned, with no pieces missing and a collection of beautiful black eyes to match the ones they sent to several other countries around the world. (I got two!)

All of these black belts proved that they were among the best in the world and I am proud to call them my friends, students, and fellow athletes.

All returned with new knowledge and stories to tell. Two returned with a new weight around their necks. Miss Maeve Brown, a farmer, rancher, rascal, 2nd degree black belt, (and my sister, so I'm double proud) brought home the bronze medal in Special Technique Breaking. That's an event where the boards are placed way up high (6.5' and up), or otherwise requiring serious acrobatics just to reach, and then the black belt has to jump and kick these improbable targets with enough precision and force to break the board. It's pretty neat.

During the trip, our family spent a lot of time admiring the lush, green fields of Italy and the even greener landscape of Switzerland, I irrevocably increased my addiction to espresso, we got to see marvels and wonders of ancient civilization that were surreal in their scope and precision, we saw Enzo Ferrari's house, the hometown of the violin, and survived a week of riding the Rome Metro Transit System.

Audra Brown won Gold in Power Breaking (again) and is now a two-time World Champion.

www.ingramcontent.com/pod-product-compliance
Lightning Source LLC
Chambersburg PA
CBHW021928190326
41519CB00009B/953